INTRODUCTION TO MODERN VIROLOGY

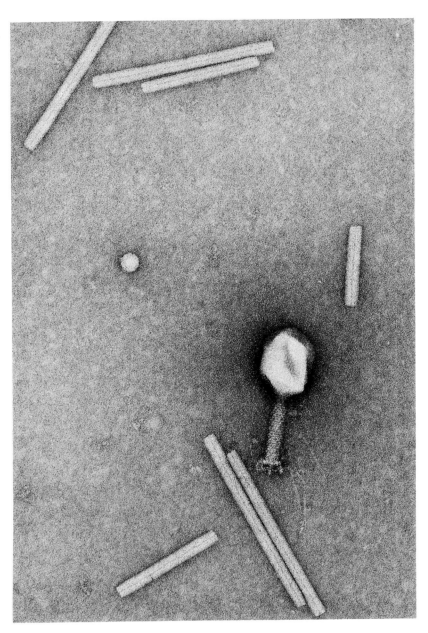

Electron micrograph showing particles of bacteriophage T4, bacteriophage φX174 and tobacco mosaic virus (courtesy Dr F. Eiserling).

Basic microbiology

EDITOR: J. F. WILKINSON

Volume 2

Introduction to modern virology

S. B. PRIMROSE B.Sc., Ph.D.
Department of Biological Sciences
University of Warwick, Coventry

N. J. DIMMOCK B.Sc., Ph.D.
Department of Biological Sciences
University of Warwick, Coventry

Second Edition

BLACKWELL SCIENTIFIC PUBLICATIONS
OXFORD LONDON EDINBURGH BOSTON MELBOURNE

© 1974, 1980
Blackwell Scientific Publications
Editorial offices:
Osney Mead, Oxford, OX2 0EL
8 John Street, London, WC1N 2ES
9 Forrest Road, Edinburgh, EH1 2QH
214 Berkeley Street, Carlton
  Victoria 3053, Australia

First published 1974
Reprinted 1978
Second edition 1980

British Library
Cataloguing in Publication Data

Primrose, S B
  Introduction to modern virology. – 2nd ed. –
  (Basic microbiology; vol. 2).
  1. Viruses
  I. Title   II. Dimmock, N J
  III. Series
  576'.64   QR360

  ISBN 0-632-00463-0

Distributed in the U.S.A. and Canada by Halsted Press, a
Division of John Wiley & Sons, Inc., New York

Set by D M B Services (Typesetting)
Cowley, Oxford,
printed in Great Britain at
The Alden Press, Oxford
and bound by
Kemp Hall Bindery, Oxford

Contents

Preface to the second edition

In the six years since the first edition of this book there have been many changes in virology. Not the least of these have been the considerable advances in animal virology and the decline of bacteriophage research. Although some of these changes are reflected in this edition our underlying aims remain the same: namely, to treat bacterial, plant and animal viruses as like entities and to use the best examples to illustrate particular principles of virology. Thus no apology is made for continued prevalence of phage examples in some sections of this edition. Some reviewers of the first edition, all animal virologists, felt that too many phage examples were used, forgetting that many ideas then current in animal virology had developed from studies with bacteriophages. The greatly increased discussion of animal viruses in this edition is not meant to appease such reviewers but is a reflection on the changes and advances which have recently taken place in animal virology. Nevertheless, in compiling this edition we have tried to accommodate those reviewers whose comments were constructively critical.

We are delighted to acknowledge the comments of R. J. Avery, A. J. Gibbs and B. D. Harrison, which have been invaluable in the preparation of the manuscript. Lastly, we are especially grateful to Debbie Bowns, our typist.

Preface to the first edition

This book has been developed from a series of ten lectures which were presented to Honours students in microbiology at the University of Edinburgh in 1971 and 1972. In these lectures I did not attempt to cover the entire field of virology. Instead, emphasis was placed on the biochemical and genetical aspects, since I was sure that the students would find this approach much more interesting than clinical details of viral diseases. Subsequent response to the lectures showed this supposition to be correct and, consequently, a similar approach has been adopted in this book. The reader will quickly realise that only a limited range of viruses is discussed; not only does this make the text easier for students, but it follows the preference of virologists for a few model systems.

Anyone who has a knowledge of virology will appreciate that research in some areas is expanding so rapidly that it is difficult to include the latest findings. However, where possible, the appropriate sections were revised shortly before submission of the manuscript.

Acknowledgement must be made of the contributions of Professor D. C. Burke and Dr J. Bennett who read most of the chapters and suggested numerous improvements of language, style and reasoning. In addition, several other colleagues criticsed individual chapters, in particular, Dr N. Dimmock (Chapter 7), Dr J. Gross (Chapter 6), and Dr R. Weiss (Chapter 11). Finally, and most important of all, I must thank my wife for her patience during the writing of this book.

1 Towards a definition of a virus

Towards the end of the last century the germ theory of disease was formulated and pathologists were confident that for each infectious disease there would be found a micro-organism which could be seen with the aid of a microscope, could be cultivated on a nutrient medium and could be retained by filters. There were, admittedly, a few organisms which were so fastidious that they could not be cultivated *in vitro* but they did satisfy the other two criteria. However a few years later, in 1892, Iwanowski was able to show that the causal agent of tobacco mosaic passes through a bacteria-proof filter and could not be seen or cultivated. Iwanowski remained unimpressed by his discovery but Beijerinck upon repeating the experiments in 1898 was convinced of the existence of a new form of infective agent which he termed *Contagium Vivum Fluidum*. In the same year Loeffler and Frosch came to the same conclusion regarding the cause of foot and mouth disease. Furthermore, because foot and mouth disease could be passed from animal to animal, with great dilution at each passage, the causative agent had to be reproducing and thus could not be a bacterial toxin. Viruses of other animals were soon discovered. Ellerman and Bang reported the cell-free transmission of chicken leukaemia in 1908 and in 1911 Rous discovered that solid tumours of chickens could be transmitted by cell-free filtrates. These were undoubtedly the first indications that viruses can cause cancer.

Although studies on bacterial viruses were later to prove the most useful for investigating the virus-host relationship they were the last to be discovered. In 1915 Twort published an account of a glassy transformation of micrococci. He had been trying to culture the smallpox agent on agar plates but the only growth obtained was that of some contaminating micrococci. Upon prolonged incubation some of the colonies took on a glassy appearance and once this occurred no bacteria could be subcultured from the affected colonies. If some of the glassy material was added to normal colonies they too took on a similar appearance even if the glassy material was first passed through very fine filters. Among the suggestions that Twort put forward to explain the phenomenon was the existence of a bacterial virus or the secretion by the bacteria of an enzyme which could lyse the producing cells. This idea of self-destruction by secreted enzymes was to prove a controversial topic for at least the next decade. In 1917 d'Hérelle observed a similar phenomenon in dysentery bacilli. He observed clear spots on lawns of dysentery bacilli and although realizing that this was not an original observation resolved to find an explanation for them. Upon noting the lysis of broth cultures of pure dysentery bacilli by filtered emulsions of faeces he immediately realized he was dealing with a bacterial virus. Since this virus was incapable of multiplying except at the expense of living bacteria he called his virus a *bacteriophage,* or *phage* for brevity.

Thus the first definition of the new type of agents, the viruses, was presented entirely in negative terms: they could not be seen or cultivated and most important of all, they were not retained by bacteria-proof filters.

The assay of viruses

The observations of d'Hérelle, however, led to the introduction of two important techniques. The first of these was the preparation of stocks of bacterial viruses by lysis of bacteria in liquid cultures. This has proved particularly valuable in modern virus research since bacteria can be grown in defined media to which radioactive precursors can be added to 'label' selected viral components. Secondly, d'Hérelle's observations provided means of assaying these invisible agents. One method was to grow a large number of identical cultures of a susceptible bacterium and to inoculate these with dilutions of the virus-containing sample. If the sample was diluted too far, none of the cultures would lyse. However, in the intermediate range of dilutions not all of the cultures lyse since not all receive a virus particle and the assay is based on this. For example, d'Hérelle noted that in 10 test cultures inoculated with a volume corresponding to 10^{-11} ml only three were lysed. Thus three cultures received one or more viable phage particles while the remaining seven received none and it can be concluded that the sample

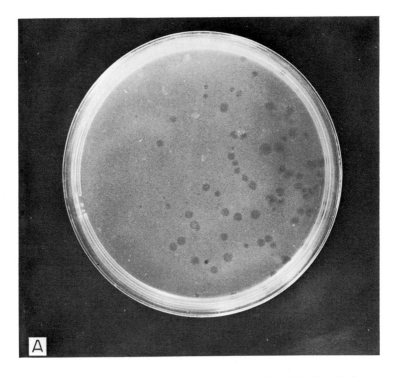

Figure 1.1 Plaques of viruses. (A) Plaques of bacteriophage Ecl on *Escherichia coli*. On opposite page: (B) Plaques of influenza virus on chick embryo fibroblasts; (C) Local lesions on a leaf of *Nicotiana* caused by tobacco mosaic virus (courtesy National Vegetable Research Station).

contained between 10^{10} and 10^{11} viable phages per ml. It is possible to apply statistical methods to end-point dilution assays of this sort and obtain more precise estimates. The other method suggested was the *plaque assay method* which is the most widely used and most useful. d'Hérelle observed that the number of clear spots or *plaques* formed on a lawn of bacteria (Fig. 1.1) was inversely proportional to the dilution of bacteriophage lysate added. Thus the *titre* of a virus-containing solution can be readily determined in terms of *plaque forming units* (p.f.u.) and if each virus particle in the preparation gives rise to a plaque then the *efficiency of plating* (e.o.p.) is unity.

Both these methods were later applied to the more difficult task of assaying plant and animal viruses. However, because of the labour, time and cost involved in providing large stocks of plants and animals, the end-point dilution assay is avoided where possible. For the assay of plant viruses a variation of the plaque-assay, the *local lesion assay* was developed by Holmes in 1929. He observed that

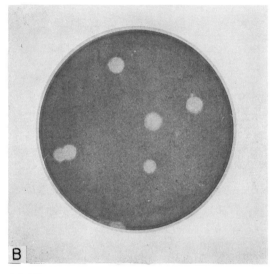

3

countable necrotic lesions were produced on leaves of *Nicotiana*, particularly *N. glutinosa*, inoculated with tobacco mosaic virus and that the number of local lesions depended on the virus content of the inoculum. Unfortunately individual plants, and even individual leaves of the same plant, produce different numbers of lesions with the same inoculum. However, the opposite halves of the same leaf give almost identical numbers of lesions and it is possible to compare the same dilutions of two virus containing samples by inoculating them on the opposite halves of the same leaf (Fig. 1.1).

A major advance in animal virology came in 1952 when Dulbecco devised a plaque assay method for animal viruses. In this case a suspension of susceptible cells, prepared by trypsinization of a suitable tissue, is placed in petri dishes or other culture vessels. The cells attach and grow across the glass surface until a monolayer of cells is formed. Once the cells are present at such a density that they come into contact with one another, growth ceases by a process called *contact inhibition*, hence the formation of a monolayer. Upon formation of the monolayer, the nutrient medium bathing the cells is removed and a suitable dilution of the virus added. After a short period of incubation to allow the virus particles to adsorb to the cells, nutrient agar is placed over the cells. After a further period of incubation ranging from 24 hours to 24 days the cells are stained by adding neutral red or some other vital dye to the agar. Living cells take up the stain but dead cells do not and the plaques are seen as unstained circular areas in the stained cell sheet (Fig. 1.1). Since tumour viruses are not cytopathic (i.e. do not kill cells) they cannot be assayed by this means. However, cells infected with these viruses do not show contact inhibition and so form 'colonies' of cells on the surface of the monolayer and this can also be used as an assay method.

The multiplication of viruses

Although methods of assaying viruses had been developed there was still considerable doubt as to the nature of viruses. d'Hérelle believed that the infecting phage particle multiplied within the bacterium and that its progeny were liberated upon lysis of the host cell. Convincing support for this hypothesis was provided by the one-step growth experiment of Ellis and Delbruck (1939). A phage preparation was mixed with a suspension of bacteria and after allowing a few minutes for the phage to adsorb, the culture was diluted 50-fold to stop further adsorption. The result obtained is shown in Fig. 1.2. After a latent period of 30 minutes in which no phage increase could be detected there was a sudden 70-fold rise in p.f.u. However, if from an infected culture very small samples are withdrawn such that each contains an average of one infected cell, then a great fluctuation in burst size is observed. Thus the burst size of 70 seen at each step in the growth curve represents the average of many different individual bursts.
Samples were withdrawn at regular intervals and assayed for phage particles.

Later Delbruck extended his studies in order to settle the controversy which surrounded the growth of phage and the dissolution of cultures. d'Hérelle believed that intracellular phage growth led to the lysis of the host cell and the liberation of virus whereas others believed that phage-induced dissolution of bacterial cultures was merely the consequence of a stimulation of lytic enzymes

4

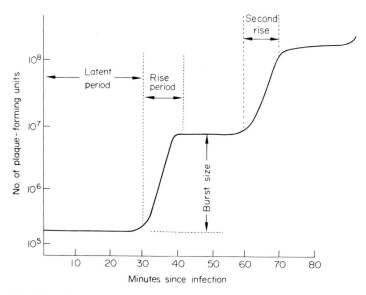

Figure 1.2 The multiplication of bacteriophages following infection of susceptible bacteria.

endogenous to the bacteria. Yet another school of thought was that phage could pass freely in and out of bacterial cells and that lysis of bacteria was a secondary phenomenon not necessarily concerned with the growth of a phage. It was Delbruck who ended the controversy by pointing out that two phenomena were involved, lysis from within and lysis from without. The type of lysis observed was dependent on the ratio of infecting phage to bacteria (*multiplicity of infection*). When the ratio of phage to bacteria is no greater than one or two, i.e. low multiplicity of infection, then the phage infect the cells, multiply and lyse the cells from within. When the multiplicity of infection is high, i.e. many more phage than bacteria, the cells lyse, there is no increase in phage titre but rather a decrease. Lysis is due to weakening of the cell wall when large numbers of phage are adsorbed.

Although the one-step growth curve demonstrated the nature and kinetics of the process by which bacterial viruses multiply within cultures of susceptible bacteria it gave no indication of the events taking place inside the cell. By using the technique of lysis from without, Doermann (1952) was able to break open the cells at selected time intervals during the growth curve. Cells of *Escherichia coli* were infected with bacteriophage T4 in the presence of cyanide. By diluting the culture no more phage can adsorb, and this also effectively removes the cyanide, thus synchronizing the intracellular events. At various times after removal of the cyanide, the cells were mixed with sufficient bacteriophage T6 to cause lysis from without. The number of T4 p.f.u. present inside the cell was determined by plating the infected suspension on a host cell which was resistant to T6. Typical results are shown in Fig. 1.3. Whereas the number of *infectious centres* (i.e. free phage + infected cells) remains constant until lysis, the number of bacteriophages inside the cell does not. Immediately after infection there is an *eclipse* period in which no phage can be detected followed by an increase in plaque

5

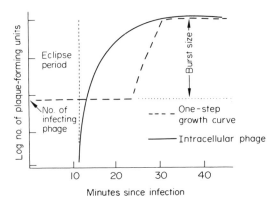

Figure 1.3 Intracellular development of bacteriophages following infection of susceptible bacteria. Note that samples taken early (before 12 minutes) contained less infectious phage than the original inoculum.

forming units until lysis occurs. It should also be noted that the kinetics of appearance of intracellular phage particles are *linear,* not exponential. This suggests that the particles are produced by assembly from component parts rather than by binary fission.

Viruses can be defined in chemical terms

The first virus was purified in 1933 by Schlessinger using differential centrifugation. This technique, which is still widely used by plant virologists, consists of several cycles of low and high speed centrifugation. The low speed centrifugation pellets the host cell debris and the high speed centrifugation pellets the virus. Chemical analysis of the purified bacteriophage showed that it consisted of approximately equal proportions of protein and deoxyribonucleic acid. A few years later in 1935 Stanley isolated tobacco mosaic virus in paracrystalline form and this crystallization of a biological material thought to be alive raised many philosophical questions about the nature of life. In 1937 Bawden and Pirie extensively purified tobacco mosaic virus and showed it to be a nucleoprotein containing RNA.

The importance of viral nucleic acid

The purification of viruses and the demonstration that they consisted of only protein and nucleic acid came at a time when the nature of the genetic material was not known. In 1949 Markham and Smith found that the spherical particles in preparations of turnip yellow mosaic virus were of two types, only one of which contained nucleic acid. Significantly, only the particles containing nucleic acid were infective. A few years later, in 1952, Hershey and Chase demonstrated the independent functions of viral protein and nucleic acid. Bacteriophage T2 was grown in *E. coli* in the presence of ^{35}S (as sulphate) to label the protein moiety, or ^{32}P (as phosphate) to label the nucleic acid. Purified labelled phage were allowed to adsorb to sensitive host cells and then subjected to the shearing forces of a Waring blender (Fig. 1.4).

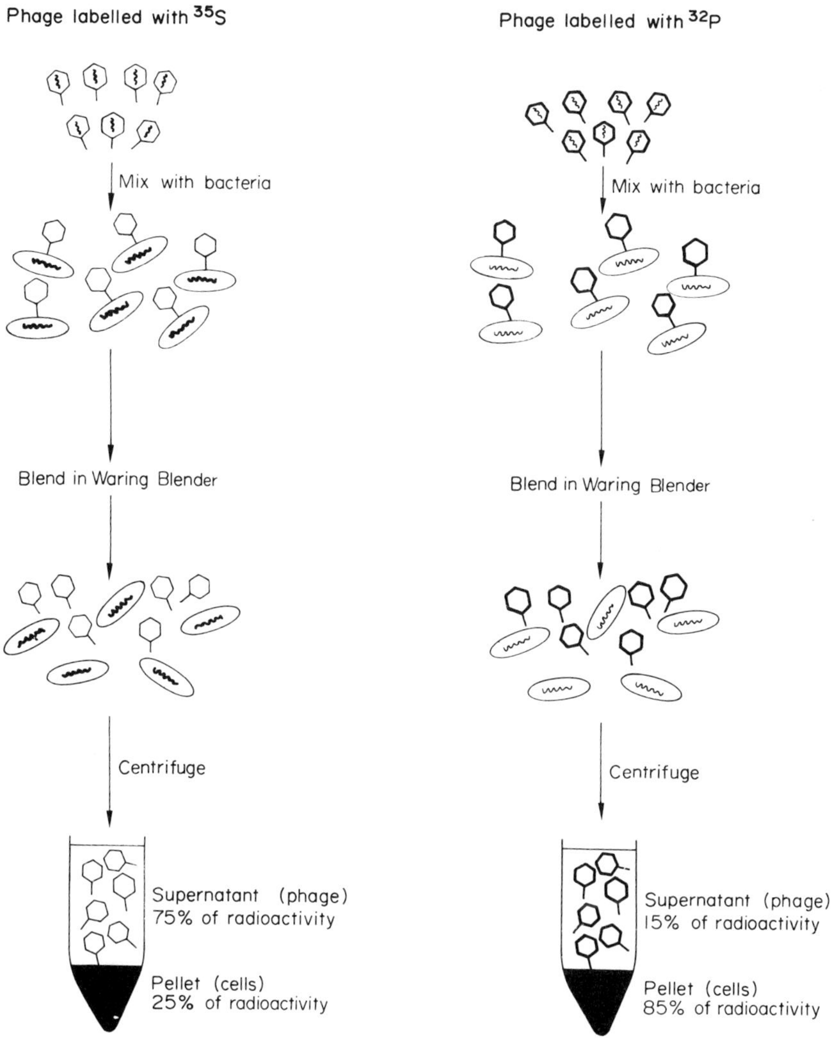

Phage labelled with ^{35}S

Phage labelled with ^{32}P

Mix with bacteria

Mix with bacteria

Blend in Waring Blender

Blend in Waring Blender

Centrifuge

Centrifuge

Supernatant (phage)
75% of radioactivity

Supernatant (phage)
15% of radioactivity

Pellet (cells)
25% of radioactivity

Pellet (cells)
85% of radioactivity

Figure 1.4 The Hershey-Chase experiment.

Treatment of the cells in this way removes any phage components attached to the outside of the cell but does not affect viability. When the cells were removed from the medium it was observed that 75% of the ^{35}S (i.e. phage protein) had been removed from the cells by blending but only 15% of the ^{32}P (i.e. phage DNA). Thus after infection the bulk of the phage protein appears to have no further function and consequently it must be the DNA that is the carrier of viral heredity. The release of the phage DNA from its protein envelope upon infection also accounts for the existence of the eclipse period during the early states of intracellular virus development since the DNA on its own cannot normally infect a cell.

7

In another classic experiment Fraenkel-Conrat and Singer (1957) were able to confirm by a different means the hereditary role of viral RNA. Their experiment was based on the earlier discovery that particles of tobacco mosaic virus can be dissociated into their protein and RNA components and then reassembled to give particles which are morphologically mature and fully infective. When two different strains (differing in the symptoms produced in the host plant) were each disassociated and the RNA of one reassociated with the protein of the other, and vice versa, the type of virus which was propagated when the resulting 'hybrid' particles were used to infect host plants was always that from which the RNA was derived (Fig. 1.5).

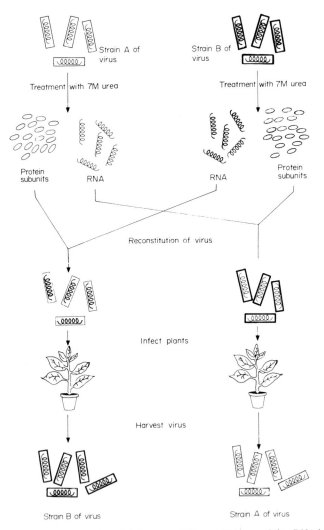

Figure 1.5 The experiment of Fraenkel-Conrat and Singer which proved that RNA is the genetic material of tobacco mosaic virus.

The ultimate proof that viral nucleic acid is the genetic material comes from numerous observations that under special circumstances purified viral nucleic acid is capable of initiating infection, albeit with a reduced efficiency. For example, in 1956 Gierer and Schramm and Fraenkel-Conrat independently showed that the purified RNA of tobacco mosaic virus can be infectious provided precautions are taken to protect it from inactivation. In fact, the causative agents of potato spindle tuber disease and citrus exocortis disease completely lack a protein component and appear to consist solely of RNA. Because they have no protein coat they cannot be called viruses and are referred to as *viroids*.

The synthesis of macromolecules in infected cells

Knowing that it is the nucleic acid which is the carrier of genetic information, and that with bacteriophages only the nucleic acid enters the cell, it is pertinent to determine the events occurring inside the cell. The discovery in 1953 by Wyatt and Cohen that the DNA of the T-even bacteriophages T2, T4 and T6 contains hydroxymethylcytosine (HMC) instead of cytosine made it possible in the same year for Hershey, Dixon and Chase to examine phage infected bacteria for the presence of phage-specific DNA at various stages of intracellular growth. DNA was extracted from T2 infected *E. coli* at different times after the onset of phage growth and analysed for its content of hydroxymethylcytosine. Analyses of this type provided an estimate of the number of phage equivalents of HMC containing DNA present at any time based on the total nucleic acid and relative HMC content of the intact T2 phage particle. The results showed that with T2, synthesis of phage DNA commences about 6 minutes after infection and then rises sharply so that by the time the first infective particles begin to appear 6 minutes later there are 50-80 phage equivalents of HMC. Thereafter the numbers of phage equivalents of DNA and of infective particles increase linearly and at the same rate up until lysis, even if lysis is delayed beyond the normal burst time. The observed kinetics of phage and DNA synthesis are important because they provide evidence not only for assembly of the virus from individual components but for regulation of development.

Hershey and his co-workers also studied the synthesis of phage protein which can be distinguished from bacterial protein by its specific interaction with phage antiserum. During infection of *E. coli* by T2 phage protein can be detected about 9 minutes after the onset of the latent period, i.e. after DNA synthesis begins, and by the time infectious particles begin to appear a few minutes later there are approximately 30-40 phage equivalents inside the cell. Whereas the synthesis of viral protein starts about 9 minutes after the onset of the latent period, it was shown by means of pulse-chase experiments that the uptake of ^{35}S into intracellular protein is constant from the beginning. A small quantity of ^{35}S (as sulphate) was added to the medium (pulse) at different times after infection and was followed shortly by a vast excess of unlabelled sulphate (chase) to stop any further incorporation of label. When the pulse was made from the 9th minute onward the label could be chased into material serologically identifiable as phage coat protein. However, if the pulse was made early in infection it could be chased into protein but although it was non-bacterial it did not react with antisera to phage structural

9

proteins. The nature of this early protein will be discussed in Chapter 8.

Being the genetic material, the nucleotide sequence in the viral nucleic acid has to be translated into proteins. The pioneering work of McQuillen, Britten and Roberts at the Carnegie Institution in Washington had shown that protein synthesis takes place on the ribosomes rather than on DNA and studies on infected cells were to solve many of the mysteries of protein synthesis. Volkin and Astrachan examined the RNA from infected and uninfected cells and compared the base ratios with the DNA from the infecting bacteriophage and from uninfected cells. From the results shown in Table 1.1 it is clear that after infection only

Table 1.1 Base Ratios of DNA and RNA from uninfected and infected Cells.

Material	Ratio of $\dfrac{\text{Adenine} + \text{Thymine/Uracil}}{\text{Guanine} + \text{Cytosine}}$
DNA from uninfected Cell	1.0
RNA from uninfected Cell	0.85
DNA from Phage	1.8
RNA from infected Cell	1.7

phage-specific RNA is synthesized and with this information Brenner, Jacob and Meselson were able to show in 1959 that this viral-specific RNA associates with pre-existing host cell ribosomes. However, before describing their experiments it is necessary to explain briefly the principles of density gradient centrifugation.

When a small amount of macromolecule (DNA, RNA, ribosome, etc.) in a concentrated solution of a heavy metal salt such as CsCl is centrifuged until equilibrium is reached, the opposing forces of sedimentation and diffusion produce a stable concentration gradient with a continuing increase in density along the direction of the centrifugal force. The macromolecules are driven by the centrifugal field into the region where the solution density is equal to their own buoyant density. At equilibrium, a single species of macromolecule is distributed over a narrow band. If several different density species of macromolecule are present each will form a band at the position where the density of the CsCl equals the buoyant density of the species (Fig. 1.6). Components banding at different places in the gradient by virtue of their differences in density may be isolated by puncturing the bottom of the centrifuge tube and collecting fractions of one or several drops in size.

Brenner *et al.* mixed ribosomes from cells grown in medium containing heavy isotopes ($^{13}C^{15}N^{32}P$) with a large excess of ribosomes from cells grown in light medium ($^{12}C^{14}N^{31}P$). The ribosome mixture was centrifuged to equilibrium on a CsCl density gradient and the fractions examined for radioactivity due to ^{32}P ('heavy' ribosomes) and for absorbance at 254 mμ ('light' ribosomes). The result obtained (Fig. 1.7) showed that ribosomes can be separated into two components designated A (more dense) and B (less dense) and that component B from 'heavy' ribosomes bands at the same place as component A from the 'light' ribosomes. If cells are pulsed with ^{14}C-uridine after infection and the ribosomes extracted and banded in CsCl, the radioactivity, which will all be in viral RNA,

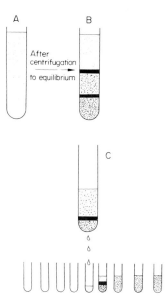

Figure 1.6 Centrifugation of macromolecules in CsCl. (A) Before centrifugation the macromolecules and the CsCl are evenly dispersed. (B) During centrifugation the CsCl forms a continuous density gradient and the macromolecules band at points where their density equals that of the CsCl. (C) The bottom of the centrifuge tube is punctured and fractions are allowed to drip into a series of tubes. These can then be assayed for radioactivity, etc.

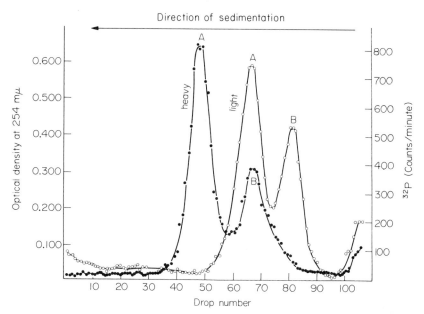

Figure 1.7 The separation of 'heavy' ($^{13}C\,^{15}N\,^{32}P$) and 'light' ($^{12}C\,^{14}N\,^{31}P$) ribosomes in a CsCl density gradient. Note that both sets of ribosomes separate into two components, A and B, and that component B of the 'heavy' ribosomes has the same density as component A of the 'light' ribosomes. (Courtesy of Dr S. Brenner and the editors of *Nature*.)

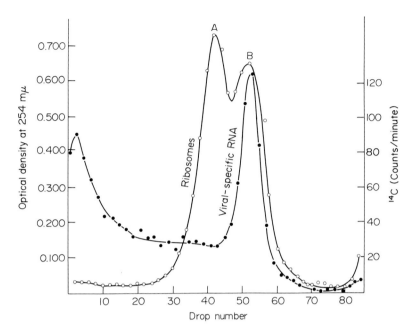

Figure 1.8 Association of viral-specific RNA with component B of ribosomes. The positions of the ribosome components are determined by absorbence and that of the RNA by radioactivity. (Courtesy of Dr S. Brenner and the editors of *Nature.*)

is found associated with component B (Fig. 1.8). A certain amount of radioactivity is found at the bottom of the gradient and this corresponds to free RNA.

Next, cells were grown in $^{15}N^{13}C$ containing medium so that they synthesized 'heavy' ribosomes, infected with phage, shifted to 'light' medium and pulsed with ^{32}P to label the viral specific RNA. These cells were then mixed with an excess of cells grown in 'light' medium and the ribosomes purified and banded in CsCl. Most of the radioactivity due to ^{32}P was found in the same position as component A of the 'light' ribosomes. Since component B of the 'heavy' ribosomes bands in the same position (Fig. 1.9) and since the virus-specified RNA normally associates with component B of the ribosomes it can be inferred that the radioactivity due to ^{32}P is bound to component B of the 'heavy' ribosomes. Since no ^{32}P is found in the position of component B of the 'light' ribosomes, the viral-specific RNA must be associated with ribosomes synthesized prior to infection.

Furthermore, when a pulse of ^{35}S is given instead of a pulse of ^{32}P this too is found associated with component B of the 'heavy' ribosomes indicating that protein synthesis is also occurring on the ribosomes.

12

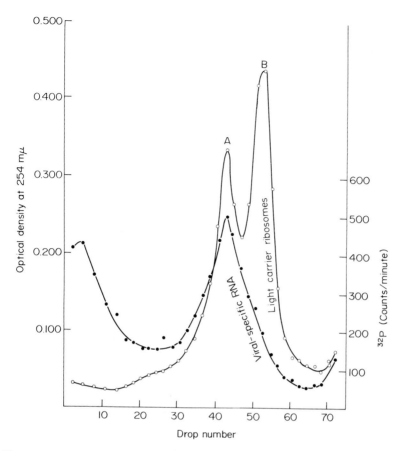

Figure 1.9 Association of viral-specific RNA with component B of the 'heavy' ribosomes. Because no radio-isotope was present during the growth in heavy medium, the position of the 'heavy' ribosomes cannot be detected directly. By adding an excess of 'light' ribosomes we can indirectly locate component B of the 'heavy' ribosomes because this has the same density as component A of the 'light' ribosomes (see Fig. 1.7). This is also the density at which the viral-specific RNA is found. (Courtesy of Dr S. Brenner and the editors of *Nature*.)

Viruses can be manipulated genetically

One of the easiest ways to understand the steps involved in a particular reaction is to isolate mutants which are unable to carry out that reaction. Like all other organisms, viruses sport mutants in the course of their growth and these mutations can affect the type of plaque formed, the range of hosts which the virus can infect or even the physico-chemical properties of the virus. There is one obvious restriction however and this is that many mutations will obviously be lethal to the virus and remain undetected. This problem was very neatly overcome in 1963 by Epstein and Edgar and their collaborators with the discovery of *conditional lethal mutants*. One class of mutants, the *temperature sensitive mutants* were able to grow at some low temperature, the *permissive* temperature, but not at some higher,

13

restrictive temperature at which normal virus could grow. The other class of conditional lethal mutants were the *amber* mutants which could not grow on a *non-permissive* host cell but could grow on a *permissive* strain. Since the first discovery of conditional lethal mutants various other types have been described. These are *ochre* mutants which are not unlike the *amber* mutants, and *cold-sensitive* mutants which are the reverse of temperature sensitive mutants.

The properties of viruses

Assuming that the features of virus growth just described for particular viruses are true of all viruses we are now in a position to compare and contrast the properties of viruses with those of their host cells. Whereas their host cells contain both types of nucleic acid, viruses only contain one type as analyses of purified viruses have shown (Schlessinger; Bawden and Pirie). However, just like their host cells, viruses have their genetic information encoded in their nucleic acid (Hershey and Chase; Fraenkel-Conrat and Singer). Another difference is that the virus is exclusively reproduced from its genetic material (Hershey and Chase; Fraenkel-Conrat and Singer) whereas the host cell is reproduced from the integrated sum of its components. Thus the virus never arises directly from a pre-existing virus whereas the cell always arises directly from a pre-existing cell. The experiments of Hershey and his collaborators showed quite clearly that the components of a virus are synthesized independently and then assembled into mature virus particles. In contrast, growth of the host cell consists of an increase in the amount of all its constituent parts during which the individuality of the cell is continuously maintained. Finally, the experiments of Brenner, Jacob and Meselson clearly showed that upon infection viruses are incapable of synthesizing ribosomes but instead depend on pre-existing host cell ribosomes for synthesis of viral proteins. These features clearly separate viruses from all other organisms, even chlamydiae which for many years were considered to be intermediate between bacteria and viruses.

PLAN OF THE REMAINDER OF THE BOOK

In the preceding discussion we have described the key experiments which led to our understanding of the nature of viruses. In addition, we have attempted to introduce the reader to the various experimental techniques used by virologists. However, animal virologists often work with animal cells in culture, rather than whole animals, and the special techniques which this involves are outlined in the ensuing appendix.

From the work of Schlessinger, Stanley and Bawden and Pirie it is clear that viruses consist of nucleic acid, the repository of the genetic information, surrounded by a protective protein coat. In Chapter 2 consideration will be given to the limited number of ways in which such protein shells can be constructed and Chapter 3 will detail the numerous structural peculiarities exhibited by viral nucleic acids. Next, our attention will be drawn to the way in which such viruses multiply inside infected cells. What mechanisms have viruses evolved to enable them to infect susceptible cells? Chapter 4, which examines this question, is unique in that it is the only chapter in which plant, animal and bacterial viruses are discussed separately. The properties of these three classes of virus are so similar, except for the penetration step, that in our opinion they do not normally warrant separate discussions. The experiments of Doermann established that viral protein and viral nucleic acid is synthesized independently and that only late in infection do the two coalesce to yield mature virus particles.

In considering the replication of viral nucleic acid we must bear in mind the numerous structural peculiarities of such molecules. The Baltimore classification which facilitates such considerations is presented in Chapter 5. The replication of viral DNA and RNA occupies the next two chapters. Not all proteins synthesized in the infected cell are structural components of the virus (Hershey *et al.*) and the molecular events controlling their synthesis are discussed in Chapter 8. Finally, this analysis of the process of infection is concluded in Chapter 9, when the formation of mature virus particles from their different components is described.

The second half of the book is devoted to the biological interactions between viruses and their hosts as distinct from the molecular aspects of replication. Chapter 10 is devoted to lysogeny, the process whereby some bacteriophages are maintained in the cell in a state of 'suspended animation'. As well as being of intrinsic interest, lysogeny provides insights towards an understanding of the interaction between viruses and animals. These are outlined in the succeeding 6 chapters. Chapter 11 is a brief account of the interaction between viruses and cultured eukaryotic cells whereas Chapter 13 is devoted to the interaction between viruses and whole animals. From the viruses point of view, one significant difference between cells in culture and whole animals is that the latter possess defence systems to modulate the effects of infection. These defence systems are described in Chapter 12 and the way man makes use of such systems in the development of vaccines is covered in Chapter 14. As a result of the widespread fear of cancer the tumour viruses are currently being subjected to intensive study and Chapter 15 summarizes the latest information in this area. Like all living organisms viruses continually evolve and the effects of this evolution on the patterns of disease in man are discussed in Chapter 16.

In any branch of science there are fashionable topics and virology is no exception. Chapter 17 contains a brief history of these trends and attempts to predict those topics which will be the focus of attention in the next few years. Finally the last chapter contains a brief description of the major groups of viruses currently recognized and should prove useful as a reference source.

APPENDIX: How to handle animal viruses

Viruses are too small to be seen except by electron microscopy and this requires concentrations in excess of 10^{11} particles, or even higher if a virus has no distinctive morphology. Therefore viruses are usually detected by other *indirect* methods.

These fall into the two categories: (1) *multiplication* in a suitable culture system and detection of the virus by the effects it causes, or (2) *serology* which makes use of the interaction between a virus and antibody directed specifically against it.

Selection of culture systems

The culture system always consists of living cells and the choice is outlined in Table 1.2. Which is used depends on the aims of the experiment. These may be divided into isolation of viruses, biochemistry of multiplication, structural studies and study of natural infections.

The investigation of any new virus starts with ways to cultivate it. There are still many which are uncultivable, particularly those occurring in the gut, but they are present in such numbers that they were actually discovered by electron microscopy. Often a virus is suspected of causing a disease. The obvious choice of experimental animal is the natural host, although for man this is ruled out by ethical considerations. Consequently organ cultures and cells must be used. Logically these should be from the natural host and obtained from those sites where the virus multiplies in the whole animal. However, often it proves that cells from unrelated animals are susceptible, e.g. human influenza viruses were first cultivated by inoculating a ferret intranasally and now grow best in embryonated chicken eggs! Frequently viruses grow poorly on initial isolation but adapt on being passed from culture to culture.

Biochemical studies of virus infections require a virus-cell system in which nearly every cell is infected. To achieve this, large numbers of infectious particles, and hence a system which will produce them, is required. Often cells which are suitable for production of virus are different from those used for the study of virus multiplication. There is little logic in choosing a cell system, only pragmatism. Cells differ greatly and different properties make one cell the choice for a particular study and unsuitable for another. Exact control of the cell's environment is needed especially for labelling with radio-isotopes since a chemically defined medium must be prepared which lacks the non-radioactive isotope. Otherwise the specific activity of the radio-isotope would be reduced to an unusable level.

The investigation of natural infections can only be done in the natural host. The nearest approximation is usually to use inbred animals which, although often not the natural host species, combine the properties of being a living animal (rather than just part of one) with the least genetic variability.

Table 1.2 How to choose a culture system for animal viruses.

Culture system	Advantages	Disadvantages
Animal	Natural infection	Cost of upkeep is expensive. Large variation between individuals even if inbred. Therefore large numbers needed. May bite!
Organ, e.g. pieces of brain, gut, trachea	Natural infection	Many cell types present.
	Fewer animals needed	'Un-natural' since isolated from regulating substances such as hormones.
	Less variation since one animal gives many organ cultures	
	Do not bite!	
Cell	Can be cloned, therefore variation between individuals is minimal	Very unnatural since cells de-differentiate when cultured.
	Best for biochemical studies as the environment can be controlled exactly and quickly.	
	May be immortal.	

Organ cultures

We shall consider here only organ cultures from the trachea. Appropriately, these were first used by an ear, nose and throat surgeon, Bertil Hoorn, who was interested in respiratory viruses. Fig. 1.10 shows the procedure used to prepare the cultures.

Ciliated cells lining the trachea continue to beat in co-ordinated waves while the tissue remains healthy. Virus multiplication causes the synchrony to be lost and eventually for the ciliated cells to detach (Fig. 1.11). Virus is also released into fluids surrounding the tissue and can be measured if appropriate assays are available.

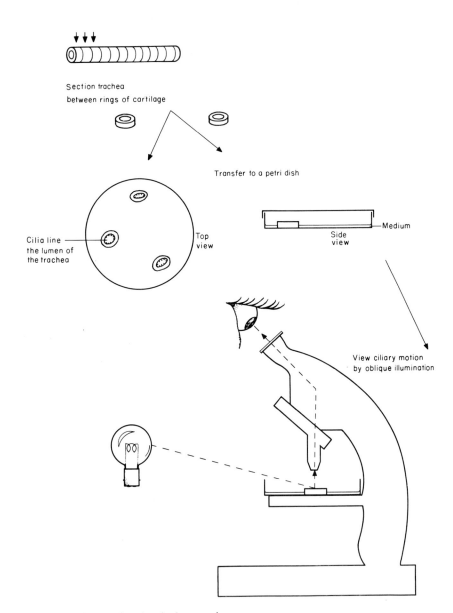

Section trachea
between rings of cartilage

Transfer to a petri dish

Cilia line
the lumen of
the trachea

Top
view

Medium

Side
view

View ciliary motion
by oblique illumination

Figure 1.10 Preparation of tracheal organ cultures.

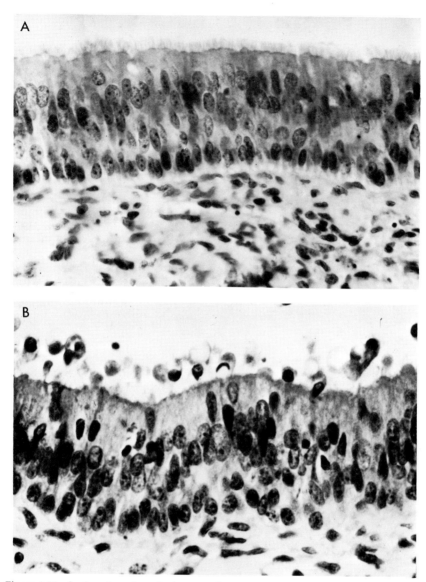

Figure 1.11 Sections through tracheal organ cultures (A) uninfected, and (B) infected with a rhinovirus for 36h. Note the disorganization of the ciliated cells after infection. (Courtesy of B. Hoorn.)

Cell cultures

Cells in culture are kept in an isotonic solution consisting of a mixture of salts in their normal physiological proportions supplemented with serum (usually 5-10%). Serum is a complex mixture of proteins and other compounds for which there is no synthetic substitute and without which mitosis does not occur. Cells in

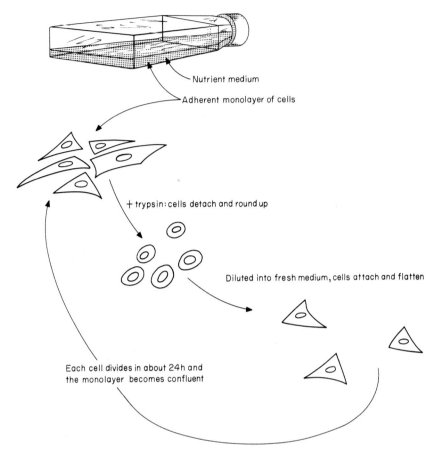

Nutrient medium

Adherent monolayer of cells

+trypsin: cells detach and round up

Diluted into fresh medium, cells attach and flatten

Each cell divides in about 24h and
the monolayer becomes confluent

Figure 1.12   Cell culture.

such a growth medium rapidly adhere to the surface of suitable glass or plastic
vessels.   Naturally all components used in cell culture are sterile and handled
under aseptic conditions to prevent the invasion of micro-organisms.   Antibiotics
have been invaluable in establishing cells in culture and routine cell culture dates
from when they appeared on the market.   Fig. 1.12 shows the principles of cell
culture.

Cultured cells are either diploid or heteroploid, having more than the diploid
number of chromosomes but not a simple multiple of it.   Diploid cells will undergo
a finite number of divisions, from around 10-100, whereas the heteroploid cells
will divide for ever.   The latter are known as *continuous cell lines* and they originate
from naturally occurring tumours or from some spontaneous event which alters
the control of division of a diploid cell.   Diploid cell lines are most easily obtained
from embryos by reducing kidneys or the whole body to a suspension of single
cells.

Haemagglutination

Certain viruses attach to receptor substances on the surface of red blood cells (rbc's) and at a high virus: cell ratio the viruses join rbc's together and the cells are agglutinated. This has nothing to do with infectivity and inactivated virus can agglutinate just as efficiently, providing its surface properties are unimpaired. A quantitative test can be devised by making dilutions of virus in a suitable tray and then adding a standard amount of rbc's to each well (Fig. 1.13). The amount of virus present is estimated as the dilution at which the virus is not

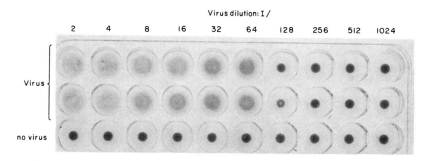

Virus dilution: I /

Figure 1.13 Haemagglutination titration, here an influenza virus is serially diluted (from left to right) in depressions in a plastic plate. 0.5% v/v red blood cells are then added and mixed with each dilution of virus. Where there is little virus, cells settle to a button (from 1/128) indistinguishable from rbc's to which no virus was added (row 3). Where sufficient virus is present (up to 1/64) cells agglutinate and settle in a diffuse pattern. (Photograph by A. S. Carver.)

sufficiently concentrated to cause agglutination. This test has the advantage of speed for it takes just 30 minutes compared with 3 days for a plaque assay. However it is insensitive since at least 10^6 p.f.u. of influenza virus are required to agglutinate rbc's.

IDENTIFICATION OF VIRUSES

Serology, or the use of antibody

Without antibody there would be no virology since there is no other way of distinguishing between morphologically similar viruses. Antibodies are proteins produced by cells of the immune system of higher vertebrates in response to foreign materials (antigens) which those cells encounter. The antibodies are secreted into the body fluids and are usually obtained from the fluid part of the blood (antiserum) which remains after clotting has removed cells and clotting proteins.

The principle of identifying infectious virus with antibody is shown in Fig. 1.14. Any method of detecting or measuring virus can be used such as inhibition of haemagglutination (Fig. 1.15) or immune electron microscopy where after reaction with specific antibody virus can be seen in aggregates. Alternatively, antibody can be employed to detect viral antigens in the infected cell. When the cell is alive, antibodies are excluded and will therefore react

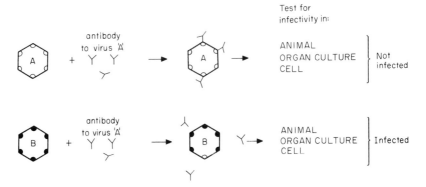

Figure 1.14 A neutralization test. Virus 'A' loses its infectivity after combining with its homologous antibody (it is neutralized) whereas virus 'B' does not. The complete test requires the reciprocal reactions.

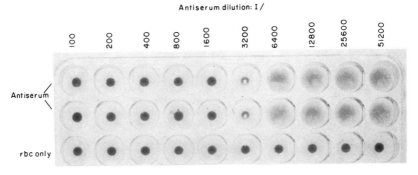

Figure 1.15 In the haemagglutination-inhibition test above, antiserum is diluted from left to right. 4 haemagglutination units (HAU) of an influenza virus are added to each well. The antibody-virus reaction goes to completion in one hour at 20°. Red blood cells are then added. In this test, haemagglutination is inhibited up to an antiserum dilution of 1/3200. (Photograph by A. S. Carver.)

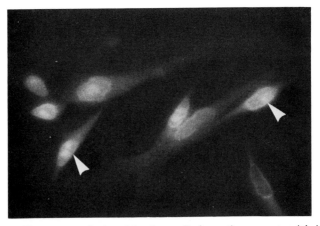

Figure 1.16 Fluorescent antibody staining (arrowed) of an antigen present mainly in the nucleus of influenza virus-infected cells.

22

only with antigens exposed on the surface of the cell. This permeability barrier is destroyed by 'fixing' the cell in acetone which enables antibody to localize antigens in the cytoplasm and nucleus. Antibodies are 'tagged' before use with a marker substance and hence can be detected *in situ*. Tags such as a fluorescent dye can be seen by UV microscopy (Fig. 1.16), an enzyme (peroxidase, phosphatase) which leaves a coloured deposit on reaction with its substrate can be seen by light microscopy, and electron dense molecules (e.g. ferritin, an iron-containing protein) are visualized by electron microscopy.

Understanding antibody reactions

The antibody-antigen reaction is so specific that it is unaffected by the presence of other proteins. Hence antibodies need not be extracted from crude serum and impure virus preparations can be used. An antibody molecule recognizes and combines with part of the antigen called the antigenic site, which is about the size of a hexapeptide or hexasaccharide, and several such antigenic sites may be presented by macromolecules forming the protein coat of a virus particle. In Chapters 12 and 13 the role of antibody in the course of virus diseases will be discussed but it is appropriate here to show how antibody affects the adsorption and penetration of virus into infected cells in culture.

Specific antibody can prevent infection by combining with virus particles in such large amounts that the particles are physically prevented from making contact with cellular receptors and adsorption cannot take place. It should be emphasized that rarely is there sufficient antibody present for this to occur and usually a virus binds but one or a few antibody molecules. These may not impede attachment or penetration to susceptible cells and may actually enhance these processes. What happens after penetration is unknown, but the viral genome is degraded intracellularly and of course no infection ensues.

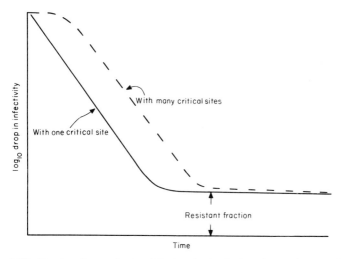

Figure 1.17 Kinetics of neutralization following the incubation of neutralizing antibody with viruses having one or many critical antigenic sites (see Fig. 1.18).

23

Formulation of a unifying hypothesis of neutralization of virus infectivity has been complicated by studies of the kinetics of the reaction (Fig. 1.17). These appear as 'single-hit' which have been interpreted as showing that one antibody molecule is all that is required for neutralizing one virus particle. However, the hypothesis is difficult to understand in terms of viruses which are made up of a number of identical repeating sub-units. For these viruses it is suggested that neutralization follows 'multi-hit' kinetics where many antibody molecules combine with a virus particle and yet do not neutralize it, but the addition of one more antibody molecule performs the *coup de grâce* and results in the loss of infectivity. In other words: virus + antibody → infectious (virus-antibody complex) + antibody → non-infectious (virus-antibody complex), i.e. neutralization.

This explanation becomes clearer in the light of the concept of 'critical' antigenic sites (Fig. 1.18). Only these are relevant to neutralization. Thus combination of one molecule of antibody to a single antigenic site of phage T4

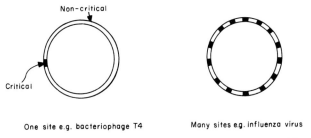

One site e.g. bacteriophage T4   Many sites e.g. influenza virus

Figure 1.18 Representation of 'critical' antigenic sites on the surface of virus particles.

located on the base plate (see Fig. 2.11 or 2.12) causes neutralization. Other sites are essential structural elements but are non-critical: thus, though antibody is raised against them and combines with them, neutralization does not occur. Other, and perhaps the majority of viruses have multiple critical sites and neutralization occurs only when sufficient of these are combined with antibody. The attachment of sub-neutralizing amounts of antibody to virus particles can be demonstrated by using radioactively labelled antibody. In the case of influenza virus, critical sites would be the haemagglutin spikes and non-critical sites would be the neuraminidase spikes since antibody to neuraminidase has no neutralizing activity.

Going back to Fig. 1.17 it can be seen that some of the infectivity is not neutralized and remains as a 'persistent fraction'. The virus particles in the persistent fraction are not genetically different but are found inside aggregates of particles which protect them from contact with antibody.

24

2 The structure of viruses

In the previous chapter we described the experiments of Hershey and Chase and Fraenkel-Conrat and Singer from which it was apparent that the ability of a virus to reproduce resides solely in its nucleic acid. Analysis of purified viruses shows that they contain 50-90% protein and since nucleic acids in solution are susceptible to shearing and degradation, we can assign a protective role to the protein component. At first sight it would appear that there is an enormous variety of ways in which the protein could be arranged round the nucleic acid in order to protect it. On the contrary however, only a limited number of designs are observed and we will now discuss briefly the limiting factors.

Viruses are constructed from sub-units

Before considering the architecture of viruses it is worth remembering that although proteins may have a regular secondary structure in the form of an α helix the tertiary structure of the protein is not symmetrical. This, of course, is a consequence of hydrogen bonding, disulphide bridges and the intrusion of proline in the secondary structure. Although we might naïvely think that the nucleic acid could be enveloped by a single, large protein molecule, this cannot be so since proteins are irregular in shape, as already stated, whereas most virus particles are regular in shape, at least when examined by electron microscopy (Fig. 2.1). However, this can also be deduced solely from considerations of the coding potential of nucleic acid molecules. A coding triplet has a molecular weight of approximately 1000 but specifies a single amino acid whose average molecular weight is about 100 daltons. Thus a nucleic acid can at best only specify one tenth of its weight of protein. Since viruses frequently contain greater than 50% protein by weight, it should be apparent that more than one identical protein of smaller molecular weight must be present.

Obviously less genetic material is required to specify a single protein molecule if it is to be used as a sub-unit, but it is not essential that the coat be constructed from identical sub-units provided the combined molecular weights of the different sub-units are sufficiently small in relation to the nucleic acid molecule which they protect. There is a further advantage in constructing a virus from sub-units, and that is greater genetic stability since reducing the size of the structural units lessens the chance of a disadvantageous mutation occurring in the gene which specifies it. If, during assembly, a rejection mechanism operates such that faulty sub-units are not included in the virus particle then an error-free structure can be constructed with the minimum of wastage.

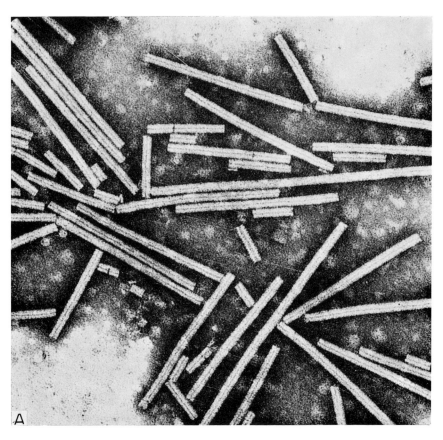

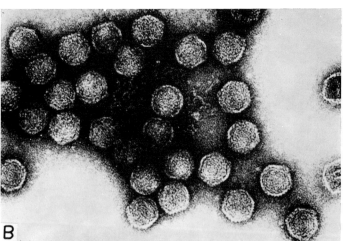

Figure 2.1 Electron micrographs of viruses showing their regular shape. (A) Tobacco mosaic virus; (B) Bacteriophage Si1.

Suspensions of pure virus can be maintained in the laboratory for long periods of time and consequently must be stable structures. The necessary physical condition for the stability of any structure is that it be in a state of minimum free energy so we can assume that the maximum number of bonds are formed between the sub-units. Since the sub-units themselves are non-symmetrical, for the maximum number of bonds to be formed they must be arranged symmetrically and there are a limited number of ways this can be done.

The structure of filamentous viruses

One of the simplest ways of symmetrically arranging non-symmetrical components is to place them round the circumference of a circle to form discs (Fig. 2.2). This gives us a two-dimensional structure. If we stack a large number of discs on top

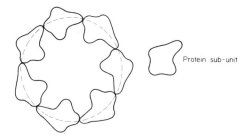

Protein sub-unit

Figure 2.2 Arrangement of identical asymmetrical components around the circumference of a circle to yield a symmetrical structure.

of one another we get a 'stacked disc' structure. Thus we can generate a symmetrical three-dimensional structure from a non-symmetrical component such as protein and still leave room for nucleic acid. Examination of published electron micrographs of viruses reveals that some of them have a tubular structure. One such virus is tobacco mosaic virus (TMV). However, close examination of TMV reveals that the sub-units are not arranged cylindrically, i.e. in rings, but helically (Fig. 2.1). There is an obvious explanation for this. Since the nucleic acid is helical in shape it could not be equivalently bonded in a stacked disc structure. However, by arranging the sub-units helically, the maximum number of bonds can still be formed and each sub-unit equivalently bonded except, of course, for those at either end. All filamentous viruses so far examined are helical rather than cylindrical and the insertion of the nucleic acid may be the factor governing this arrangement.

The structure of spherical viruses

A second way of constructing a symmetrical particle would be to arrange the sub-units around the vertices or faces of an object with cubic symmetry, e.g. tetrahedron, cube, octahedron, dodecahedron (constructed from 12 regular pentagons) or icosahedron (constructed from 20 equilateral triangles). Figure 2.3 shows possible arrangements for objects with triangular and square faces. Multiplying the number of sub-units per face by the number of faces gives the

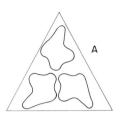

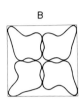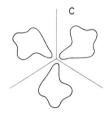

Figure 2.3 Symmetrical arrangement of identical asymmetrical sub-units by placing them on the faces of objects with cubic symmetry. (A) Asymmetrical sub-units located at vertices of each triangular facet. (B) Asymmetrical sub-units placed at vertices of each square facet. (C) Arrangement of asymmetrical sub-units at each corner of a square with face represented in (B).

minimum number of sub-units which can be arranged around such an object. For example, for a tetrahedron it is 12 sub-units, for a cube or octahedron it is 24 sub-units and for a dodecahedron or icosahedron it is 60 sub-units. Although it may not be immediately apparent, these represent the few ways in which an asymmetrical object can be placed symmetrically on the surface of a sphere. (The reader may care to check by using a ball and sticking on bits of paper of the shape shown in Fig. 2.3!) Examination of electron micrographs reveals that many viruses are spherical in outline but actually have icosahedral symmetry rather than octahedral, tetrahedral or cuboidal symmetry. There are two possible reasons for the selection of icosahedral symmetry over the others. First, since it requires a greater number of sub-units to provide a sphere of the same volume, the size of the sub-units can be smaller thus economizing on genetic information. Secondly, there appear to be physical restraints which prevent the tight packing of sub-units required by tetrahedral and octahedral symmetry.

The situation is more complex than outlined above since many viruses possess more than 60 sub-units. If 60n sub-units are put on the surface of a sphere, one solution is to arrange them in n sets of 60 units, but the members of one set would not be equivalently related to those in another set. For example, consider the arrangement of the sub-units in Fig. 2.4. If all the sub-units, represented by open and closed circles, are identical then those represented by closed circles are equivalently related as are those represented by open circles. However, open circle units do not have the same spatial arrangement of neighbours as closed circle units and so cannot be equivalently related. Of course, if the structure were built out of n different sub-units there would be no conceptional difficulty and, indeed, no problem. However, accepting the restriction that we must build the structure out of identical sub-units, how can we regularly arrange

28

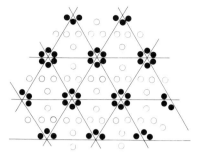

Figure 2.4 Spatial arrangement of two identical sets of sub-units. Note that any member of the set represented by closed circles does not have the same neighbours as a member of the other set represented by open circles.

Figure 2.5 An example of a geodesic dome—the United States Pavilion at Expo '67 in Montreal (courtesy of the United States Information Service).

greater than 60 asymmetrical sub-units? The solution to the problem was inspired by the geodesic domes constructed by Buckminster Fuller (Fig. 2.5). Fuller's dome designs involve the sub-division of the surface of a sphere into triangular facets which are arranged with icosahedral symmetry. The device of triangulating the sphere represents the optimum design for a closed shell built of regularly bonded identical sub-units. No other sub-division of a closed surface can give a comparable degree of equivalence. Thus, this is a minimum energy structure and hence a further reason for the preponderance of icosahedral viruses.

The triangulation of spheres

It is possible to enumerate all the ways in which this subdivision can be carried out but before doing so let us consider one simple example. If we start with an icosahedron and arrange the sub-units around the vertices there will be 12 groups of 5 sub-units (Fig. 2.6). Now we can subdivide each triangular face into four smaller and identical equilateral triangles and incorporate 240 sub-units at the

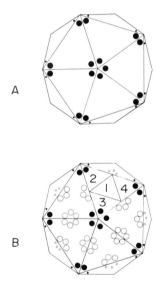

Figure 2.6 Arrangement of 60n identical sub-units on the surface of an icosahedron. (A) n = 1 and the 60 sub-units are distributed such that there is one sub-unit at the vertices of each triangular facet. Note that each sub-unit has the same arrangement of neighbours and so all the sub-units are equivalently related. (B) n = 4. Each triangular facet is divided into four smaller, but identical equilateral triangles and a sub-unit is again located at each vertex. Note that in contrast to the arrangement shown in Fig. 2.4, each sub-unit, whether represented by an open or closed circle, has the identical arrangement of neighbours. However, since some sub-units are arranged in pentamers and others in hexamers, the members of each set are only 'quasi-equivalently' related.

vertices of those smaller triangles. At the vertices of the original icosahedron there will be rings of 5 sub-units called *pentamers* (solid circles). However, at all the other vertices there will be rings of 6 sub-units, called *hexamers* (open circles). Since some of the sub-units are arranged as pentamers and others as hexamers it should be apparent that they cannot be equivalently related, hence the use of the term *quasi-equivalence,* but this still represents the minimum energy shape. Similar patterns can be obtained by triangulating the pentameric faces of a dodecahedron.

The ways in which each triangular face of the icosahedron can be subdivided into smaller, identical equilateral triangles are governed by the laws of solid geometry. These can be calculated from the expression $T = Pf^2$ where T, the *triangulation number* = the number of smaller, identical equilateral triangles, f may = 1, 2, 3, 4 etc. and P is given by the expression $h^2 + hk + k^2$, h and k being any pair of integers without common factors.

For viruses so far examined the values of P which have been found are 1 ($h = 1, k = 0$), 3 ($h = 1, k = 1$) and 7 ($h = 1, k = 2$). Representative values of T are shown in Table 2.1. Once the number of triangular subdivisions is known, the total number of sub-units can easily be determined since it is equal to 60T.

Table 2.1 Values of capsid parameters in a number of icosahedral viruses. The value of T was obtained from examination of electron micrographs thus enabling the values of P and f to be calculated.

P	f	T ($= Pf^2$)	No of sub-units (60 T)	Example
1	1	1	60	Satellite tobacco necrosis virus
1	2	4	240	Inner shell of reovirus
1	3	9	540	Outer shell of reovirus
1	4	16	960	Herpesviruses
1	5	25	1500	Adenoviruses
3	1	3	180	Turnip yellow mosaic virus
				Turnip crinkle virus
				Poliovirus
7	1	7	420	Human and rabbit papilloma viruses

The morphological units seen by electron microscopy are called capsomeres and *the number of these need not be the same as the number of protein sub-units.* The numbers of morphological units seen will depend on the size and physical packing of the sub-units and on the resolution of electron micrographs. Although the formation of shells from identical sub-units does not necessarily require clustering of these units, clustering is favoured since it maximizes interaction between the sub-units resulting in increased stability. Three types of clustering are possible and these are exemplified by poliovirus, turnip crinkle virus (TCV) and turnip yellow mosaic virus (TYMV) all of which have 180 sub-units (T = 3). In polio- virus the sub-units are clustered near the centres of the triangles giving rise to 60 morphological units. In TCV the sub-units are clustered near the centres of the edges giving rise to 90 morphological units. In the case of TYMV the sub- units are clustered into 20 hexamers and 12 pentamers (Fig. 2.7). One conse- quence of this clustering of sub-units is that the bonds between sub-units within a capsomere are stronger than those between sub-units in different capsomeres for the capsid of some viruses disintegrates into individual capsomeres during purification.

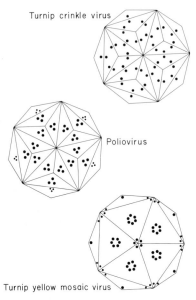

Turnip crinkle virus

Poliovirus

Turnip yellow mosaic virus

Figure 2.7 Different ways of clustering 180 sub-units as exemplified by poliovirus, turnip crinkle virus and turnip yellow mosaic virus.

Viruses with more than one kind of sub-unit

Like the viruses just described, the RNA bacteriophage R17 is constructed from 180 identical protein sub-units. However, R17 also contains one molecule per virion of a second protein, the A protein, to which the RNA is attached. Since the A protein is also involved in attachment of the virus to susceptible bacteria (Chapter 4) it must be located on the viral surface. If R17 were constructed like TYMV from 32 capsomeres then placement of the A protein within the capsid would be difficult. The 32 capsomere model has an entire shell of protein surrounding the RNA so that the A protein would have to substitute for part of the structural coat protein in order to be external enough to function during the attachment and penetration reactions and at the same time remain available for bonding to the RNA which is located internally. The A protein has a mol. wt. of 40 000 daltons and to fit this into a 32 capsomere particle would require replacement of some coat protein molecules without upsetting the quasi-equivalent bonding among sub-units. The A protein is too small for one to readily envisage how it could entirely replace a pentamer (mol. wt. 65 750 daltons) or a hexamer (79 500 daltons) and the possibility of it substituting for part of these structures is unlikely.

The 180 coat protein sub-units could be arranged as 60 trimers and the mol. wt. of each trimer (39 750 daltons) would be close to that of the A protein (40 000 daltons). Thus the A protein could substitute for a trimer of coat proteins but by doing so it would have to fulfil the bonding requirements of the missing trimer plus bond to the RNA and be capable of interacting with the attachment site on the cell.

The most satisfactory model which has been proposed so far is one in which the sub-units are arranged as nonagons, one nonagon being located on each face of the icosahedron (Fig. 2.8). The A protein could then be located at a

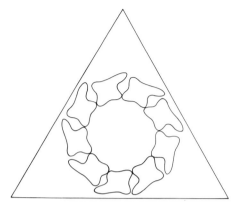

Figure 2.8   The arrangement of the protein sub-units of R17 into nonagons.

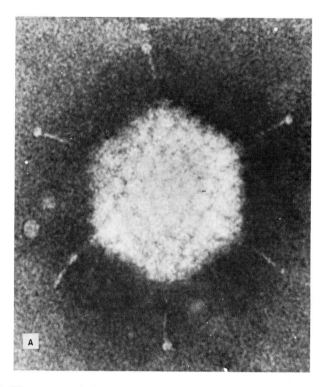

Figure 2.9   The structure of adenoviruses. (A) Electron micrograph of adenovirus (courtesy of Dr N. Wrigley). (B) Model of adenovirus to show arrangement of the capsomeres (courtesy of Dr N. Wrigley). (C) Schematic diagram to show the arrangement of the sub-units on one face of the icosahedron. Note the subdivision of the face into 25 smaller equilateral triangles.

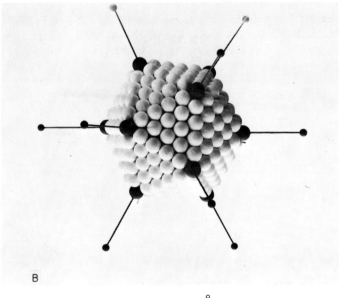

B

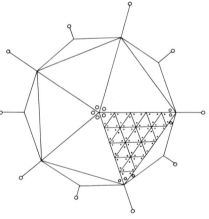

C

vertex of the icosahedron and would be surrounded by 5 nonagons, each on an adjacent face of the icosahedron. This model has the advantage over the other models in that it does not require the A protein to take the place of some of the coat protein sub-units and electron micrographs of the virus suggest that it is probably correct.

Careful examination of electron micrographs of adenoviruses (T = 25) shows that the 1500 sub-units are arranged into 240 hexamers and 12 pentamers and that there is a fibre projecting from each vertex of the virus (Fig. 2.9). It can be shown serologically that the fibres, the pentamers and the hexamers are constructed from different proteins. We are thus faced with the problem of arranging not one, but three different proteins in a regular fashion while adhering to the design principles outlined above. This can be achieved by arranging the pentamers

34

and the fibres at the vertices of the icosahedron and the hexamers on the faces of the icosahedron (Fig. 2.9). Bacteriophage ϕX174, which has much shorter spikes than adenoviruses, is probably constructed in a similar fashion.

A different structural arrangement is found in another class of spherical viruses, the reoviruses. Here the capsid is constructed from 8 different proteins but these are arranged in two layers, both of which have icosahedral symmetry. Three of the proteins are arranged in the outer shell and the remaining 5 proteins in the inner shell. The exact arrangement of the different proteins in the two shells is not yet clear but it is known that there are spikes at the icosahedral vertices of the inner shell.

ENVELOPED VIRUSES

Many of the larger animal viruses, and a few plant and bacterial viruses are enveloped by a 75 Å thick membrane layer. This envelope, which is largely derived from host-cell membranes, can be disrupted by treatment with ether or detergents and this destroys the infectivity of the virus. Such viruses are sometimes referred to as ether-sensitive viruses.

The influenza viruses

One of the best studied groups of enveloped viruses is the influenza viruses. Although these viruses are normally described as pleomorphic (Fig. 2.10) it should be noted that when first isolated from animals they are filamentous in shape. They only assume their familiar pleomorphic appearance upon passage through embryonated eggs. In electron micrographs (Fig. 2.10) a large number of surface protrusions, or spikes, projecting about 100 Å from the viral envelope can generally be observed. These spikes which have an overall length of 140 Å are embedded in the membrane like the glycoproteins of the normal cell-membrane. The spike layer consists of virus-specified glycoproteins and harbours its haemagglutinating and neuraminidase activities (Chapter 4). These reside in morphologically different spikes (Fig. 2.10).

Internal to the envelope is a layer of protein called the M protein (membrane or matrix protein). Finally, inside the M protein layer, and distinctly separated from it, is the internal nucleoprotein component; this consists of a flexible rod of RNA and protein arranged in a twisted hairpin structure. Thus influenza virus codes for at least 4 major proteins which are assembled into a virion in a variety of structural strategies.

The Rhabdoviruses ('bullet'-shaped viruses)

The viruses of this large group have as principal morphological characteristic a bullet or bacillus shape (Fig. 2.10) and often display very regular cross-structures which represent the internal helical nucleoprotein structure. There has been some discussion as to whether all the viruses in this group may in reality be bacilliform and the bullet shape an artefact produced by the negative staining procedure. Whereas this may indeed be the case with some of the plant rhabdoviruses, only

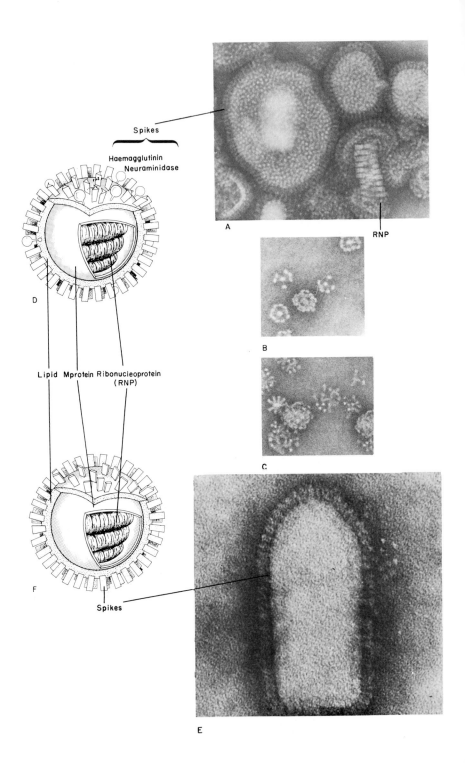

Spikes

Haemagglutinin
Neuraminidase

A

RNP

D

Lipid Mprotein Ribonucleoprotein
(RNP)

B

C

F

Spikes

E

the bullet shape is seen with animal rhabdoviruses fixed *in situ* in infected cells. Like the influenza viruses, rhabdoviruses are covered with a layer of regularly arranged spikes and have an underlying membrane-associated protein (matrix protein). To account for the unusually rigid structure of the rhabdoviruses, compared with the influenza viruses, it has been suggested that there could be a direct geometrical relationship between the matrix protein sub-units and the protein sub-units of the nucleocapsid. Otherwise the influenza and rhabdoviruses have a very similar structure.

Enveloped Icosahedral viruses

Whereas the influenza and rhabdoviruses have a helical nucleocapsid, some other enveloped bacterial and animal viruses have an icosahedral nucleocapsid (see Table 2.2). An even more complex arrangement is exhibited by the RNA

Table 2.2 Distribution of the various types of virus architecture amongst viruses of bacteria, plants and animals.

VIRUS ARCHITECTURE	DISTRIBUTION AMONGST VIRUSES OF		
	BACTERIA	PLANTS	ANIMALS
HELICAL	Relatively rare. All examples so far are male-specific	Very common	Not known
ENVELOPED HELICAL	Not known	Not common. So far only found in plant rhabdoviruses	Common
ICOSAHEDRAL	Relatively rare. Best examples are the ϕX174-related phages and the RNA-containing male-specific phages of *Escherichia coli*	Common	Common
ENVELOPED ICOSAHEDRAL	Rare. Best example PM2	Rare. Tomato spotted wilt virus	Common
ENVELOPED HELICAL + ICOSAHEDRAL	Not known	Not known	Found in oncogenic RNA viruses
HEAD + TAIL	Most commonly found	Not known	Not known

tumour viruses. These have a super-coiled nucleoprotein in the shape of a hollow sphere surrounded by an icosahedral shell which in turn is surrounded by the envelope. Once again the envelope, which is derived from the cell-membrane, is modified by the insertion of virus-specific glycoprotein.

Figure 2.10 Comparative structure of a myxovirus (influenza) and a rhabdovirus (vesicular stomatitis virus). Note that although the electron microscopic appearance is different the structural details are very similar. (A) Electron micrograph of influenza virus showing the internal ribonucleoprotein and the surface spikes. (B) Aggregates of purified neuraminidase. (C) Aggregates of purified haemagglutinin. Note the triangular shape of the spikes when viewed 'end-on'. (D) Schematic representation of the structure of influenza virus. (E) Electron micrograph of vesicular stomatitis virus. (F) Schematic representation of the structure of vesicular stomatitis virus. (Electron micrographs courtesy of N. Wrigley and Chris Smale.)

Viruses built on this architectural principle (Fig. 2.11) have thus far only been found among the bacteriophages and it seems reasonable to speculate that this is

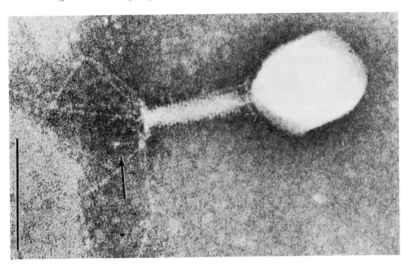

Figure 2.11 Electron micrograph of bacteriophage T2 (courtesy of Dr L. Simon). Six long tail fibres are evident. Tail pins cannot be seen but a short fibre (indicated by the arrow) can be seen. The bar is 1000Å.

connected with the way in which bacterial viruses infect susceptible cells (Chapter 4). The number of different bacteriophages with a head-tail type of architecture is very large and they can be subdivided into those with short tails, those with long non-contractile tails and those with complex contractile tails (see p. 237). A number of other structures such as base-plates, collars etc. may also be present (Fig. 2.12). Despite their more complex structure the design principles involved

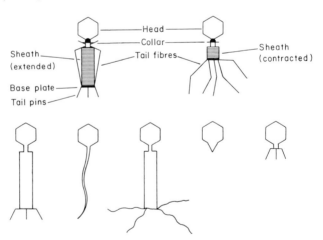

Figure 2.12 Schematic representation of the structures of some tailed bacteriophages.

are identical to those outlined earlier for the viruses of simpler architecture. Heads usually possess octahedral or icosahedral symmetry whereas tails usually have helical symmetry. All other structures such as base-plates, when present, also possess some form of symmetry, usually sixfold symmetry.

OCCURRENCE OF DIFFERENT VIRUS MORPHOLOGIES

The different virus morphologies discussed above do not occur with equal frequency among bacterial, plant and animal viruses. Information on the frequency with which they occur is summarized in Table 2.2.

THE PRINCIPLE OF SELF-ASSEMBLY

If viruses are constructed according to principles of efficient design they should be able to assemble themselves without any external organizer due to the formation of a large number of weak bonds when they are brought into the right configuration by random thermal movements. That this is indeed the case has been best shown by TMV. Treatment of TMV with 7M urea results in degradation of the virus into RNA and protein sub-units. If the urea is removed by dialysis and the RNA and protein incubated together in a suitable buffer, the polymers spontaneously aggregate to give infectious virus. If, however, the RNA is omitted during the repolymerization step, protein rods are obtained but the sub-units are arranged in a stacked disc structure rather than in a helical structure. Furthermore, these rods are much less stable than native or reconstituted virus indicating the important contribution made by the RNA to the stability of the virus. The icosahedral cowpea chlorotic mottle virus (a bromovirus) has also been reconstituted. However, when the experimental conditions are varied it is possible to get tubular structures and other variations indicating that it is the packing properties of the protein which determine the structure of the particle.

Self-assembly is economical to the virus in that it requires no specific genetic information and it may also have the advantage of providing a built-in rejection mechanism for any faulty sub-units which may be produced during replication. However, with some bacteriophages such as T4 there is evidence for genetic control of assembly and we will return to this topic in Chapter 9.

3 Viral nucleic acids

The nucleic acid of a virus contains within itself both the specific information and the operational potential such that upon entering a susceptible cell it can subvert the biosynthetic machinery of that cell and redirect it towards the specific production of viral particles. Viral nucleic acids display a remarkable array of structural and compositional varieties and since any peculiarities of the nucleic acid must have a bearing on the process of replication it is proposed to discuss some of them in detail.

The physical chemistry of the nucleic acids

Nucleic acids contain the nucleosides: adenosine, guanosine, cytidine and either uridine or thymidine. In double-stranded molecules base pairing occurs between guanosine and cytidine and between adenosine and either uridine or thymidine. It is possible to denature these double stranded molecules, i.e. separate the two strands, by heating or by alkali treatment. When heat is applied to nucleic acid solutions, the temperature at which 50% denaturation occurs (Tm) is most easily measured by following the U.V. absorbance of the sample as the temperature is increased, a sudden increase in absorbance indicating strand separation (denaturation or melting) (Fig. 3.1).

The temperature at which melting occurs is dependent upon ionic strength and on the guanine plus cytosine (% GC) content of the nucleic acid. For DNA

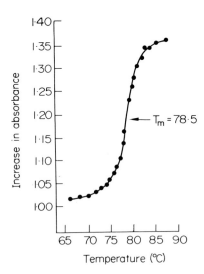

Figure 3.1 Increase in ultraviolet absorbance of DNA with increase in temperature.

in standard saline citrate (0.15M NaCl + 0.015M Na citrate) the relationship is given by:

$$\% \ GC \ = \ 2.44 \ (Tm\text{-}69.3)$$

If denatured DNA is incubated at a temperature 25° below Tm the two strands will slowly anneal but it is usually impossible to get complete annealing since a certain amount of mispairing occurs. If RNA complementary to either strand of DNA is added to the annealing mixture it is possible to obtain RNA-DNA hybrids.

As discussed in Chapter 1, it is possible to generate gradients of heavy metal salts such as CsCl by high-speed centrifugation. When the five possible classes of nucleic acid (double- and single-stranded DNA, double- and single-stranded RNA, and DNA-RNA hybrids) are centrifuged in these gradients they are all found to have characteristic buoyant densities. For double-stranded DNA the buoyant density (ρ) is dependent on the guanine plus cytosine content and the two are related as follows:

$$\% \ GC \ = \ \frac{\rho - 1.660}{0.098}$$

For single-stranded DNA the buoyant density is related to base composition by the expression

$$\rho = 1.6267A + 1.758G + 1.768C + 1.7424T$$

where A, G, C and T represent the fractions of adenine, guanine, cytosine and thymine of that DNA. When double-stranded DNA is placed in alkaline CsCl (pH 12.4) the DNA is immediately denatured and under these conditions the buoyant density of the now single-stranded DNA depends on the guanine and/or thymine content due to the titration of protons in these bases by Cs^+ ions.

It is possible to introduce a density label into DNA by growing the DNA source (phage, cell, etc.) in a medium containing either heavy isotopes (^{15}N, ^{18}O, ^{2}H) or the thymine analogue 5-bromouracil (Fig. 3.2). DNA which is given

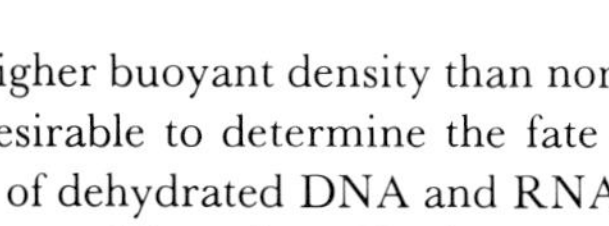

Thymine   5-bromouracil

Figure 3.2   Structure of the pyrimidine base, thymine, and its analogue, 5-bromouracil.

such a label will have a higher buoyant density than normal DNA and this can be very useful when it is desirable to determine the fate of a particular molecule.

The buoyant density of dehydrated DNA and RNA is greater than 2g/cc but in the presence of heavy metal ions the DNA is extensively hydrated leading to lower observed buoyant densities. However, the degree of hydration differs with the different salts and hence the buoyant density also varies (Table 3.1). In CsCl the buoyant density of RNA is approximately 1.9 and densities as high as this are only produced close to the saturation point of aqueous CsCl solutions.

Consequently most RNA is centrifuged to equilibrium in Cs_2SO_4 in which its buoyant density is much lower. However, CsCl is preferred for DNA since density is not a linear function of base composition in Cs_2SO_4.

Table 3.1 Buoyant density of DNA in solutions of different caesium salts.

Caesium salt used in gradient formation	Approx. buoyant density for DNA with 50% GC content
Cs Acetate	1.95 g/cc
Cs Formate	1.75 g/cc
Cs Chloride	1.70 g/cc
Cs Bromide	1.63 g/cc
Cs Iodide	1.55 g/cc
Cs Sulphate	1.45 g/cc

Analysis on sucrose density gradients (velocity or rate-zonal centrifugation)

Whereas the separation of nucleic acids in CsCl gradients is a function of base composition, their separation in sucrose density gradients is a function of size and shape. Unlike CsCl gradients, sucrose gradients are not self-generating in centrifugal fields and so must be preformed. There are several ways of doing this but the commonest method is the use of a gradient-maker (Fig. 3.3). This consists of two reservoirs joined by a capillary tube which is fitted with a valve. In addition, one of the reservoirs has an outlet tube and contains a stirring device. If a 5-20% sucrose gradient is desired, 20% sucrose is put into the mixing reservoir and 5% sucrose in the other. Stirring is begun and the exit line opened. As 20% sucrose leaves the mixing reservoir the valve separating the two reservoirs is opened to allow 5% sucrose to flow in and gradually dilute the 20% sucrose. A small volume of macromolecule-containing solution is then gently layered on top of the gradient. Upon application of a centrifugal field, the macromolecules sediment through the solution at a rate dependent on their size and shape. The rate of sedimentation is usually expressed as the sedimentation coefficient, S:

$$S = \frac{dx/dt}{\omega^2 x}$$

where x is the distance from the centre of rotation, ω the angular velocity in radians per second, and t the time in seconds. Nucleic acids have sedimentation coefficients in the range between 1 and 100×10^{-13} seconds. A sedimentation coefficient of 1×10^{-13} seconds is called a Svedberg and is abbreviated S. Thus a sedmentation coefficient of 42×10^{-13} seconds would be denoted 42S.

There is no theoretical reason why macromolecules cannot be separated by sedimentation through a column of water instead of sucrose. Practically, however, good separations are hard to achieve in this way because the slightest disturbance of the centrifuge tube will cause remixing: the presence of a gradient of sucrose minimizes such disturbances. In addition, the increasing concentration of sucrose counteracts the increasing centrifugal force imposed on nucleic acids as they move further from the centre of rotation; in this way the rate of sedimentation is kept constant.

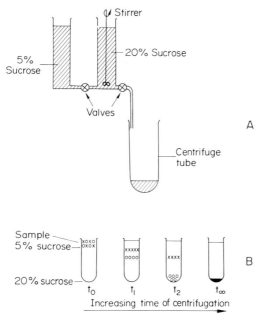

Figure 3.3 Preparation and use of sucrose density gradients. (A) Apparatus used for the preparation of sucrose density gradients (see text for details). (B) Separation of two components in a sucrose density gradient. At time t_0 the mixture of two components is layered on top of the gradient. After centrifugation for time t_1 the two components have separated and by time t_2 the faster of the components has pelleted. Note that if centrifugation is continued indefinitely both components will be pelleted and separation will not be achieved.

Electrophoresis of nucleic acids

For the separation of different RNA molecules, the use of sucrose gradients has largely been superceded by zone electrophoresis in polyacrylamide gels. In this method the centrifugal field is replaced by an electrical field and separation is again based on size and shape. Polyacrylamide gels are hydrated and porous but are mechanically rigid and so function in an analogous manner to sucrose by preventing convectional and vibrational disturbances. DNA molecules, particularly those smaller than 10^7 daltons mol. wt., can also be separated by electrophoresis (Fig. 3.4) but agarose is the material of choice for gel construction.

Armed with this brief introduction to nucleic acids we are now in a position to examine some of the structural peculiarities of those isolated from viruses.

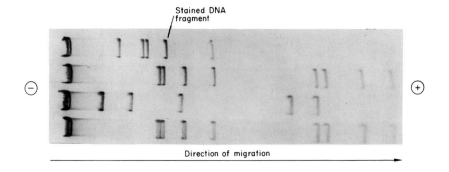

Stained DNA fragment

Direction of migration

Figure 3.4 Separation of DNA molecules of different sizes by electrophoresis in an agarose gel. DNA from bacteriophage λ was cleaved with different restriction endonucleases (see page 95) and after electrophoresis the DNA was 'stained' by immersing the gel in ethidium bromide. (Photograph courtesy of Professor K. Murray.)

TYPES OF NUCLEIC ACIDS FOUND IN VIRUSES

There are four possible kinds of viral nucleic acid: single-stranded DNA, single-stranded RNA, double-stranded DNA and double-stranded RNA. Their distribution among the different types of virus is shown in Table 3.2. The type of nucleic acid present in the particle can be determined by staining with acridine orange (Table 3.3) but it is more satisfactory to extract the nucleic acid and determine its base composition, nuclease sensitivity, buoyant density, etc. Single-stranded nucleic acids are detected by the absence of a sharp melting profile upon heating and the non-equivalence of the molar proportions of adenine and thymine (or uracil) or guanine and cytosine.

Table 3.2 Distribution of the four basic types of nucleic acid among viruses of bacteria, plants and animals.

| | DISTRIBUTION | | |
TYPE OF NUCLEIC ACID	BACTERIAL VIRUSES	PLANT VIRUSES	ANIMAL VIRUSES
Single-stranded DNA	Not very common. Best examples φX174 and filamentous phages	Rare. Only found in geminiviruses	Common
Double-stranded DNA	Most common	Rare. Only found in caulimoviruses	Common
Single-stranded RNA	Not very common. Best examples are the male-specific coliphages	Most common type found.	Common
Double-stranded RNA	Rare. Only found in cystoviruses	Only found in Reoviridae	Common

44

Table 3.3 Identification of nucleic acid type by acridine orange staining. A pure virus suspension with no free nucleic acids is dried onto a glass slide, fixed with Carnoy's fluid and stained with acridine orange. After shaking in Na_2HPO_4 the slides are examined under short wave ultraviolet light. To determine whether the particles contain DNA or RNA, the slide is treated with molybdic acid solution.

Type of nucleic acid	TREATMENT	
	Na_2HPO_4	Molybdic acid
Double-stranded DNA	Green	Green
Double-stranded RNA	Green	Green fades
Single-stranded DNA	Red	Paler green
Single-stranded RNA	Red	Paler red

Unusual bases

The DNA from certain bacteriophages is known to contain unusual bases. Examples are the replacement of thymine by uracil[†] (bacteriophage PBS1) or hydroxymethyluracil (bacteriophage SP8) and the replacement of cytosine by hydroxymethylcytosine (bacteriophages T2, T4, T6). In the T-even phage series the hydroxymethylcytosine may be further substituted with glucose or gentiobiose. Since these bases do not appear in the DNA from uninfected host cells the information specifying their synthesis must be carried by the virus. Like the host cell DNA, most viral DNA molecules are also partially methylated and subtle differences in methylation may help enzymes to distinguish between host and viral DNA. Since the equations relating base composition to buoyant density and Tm no longer hold if the DNA is modified in any way, the presence of such bases can be determined by measuring the latter two parameters. If these two methods yield significantly different values for the base composition it is a good indication of the presence of modified bases (Table 3.4).

Table 3.4 Effect of unusual bases on melting temperature and buoyant density of DNA.

Phage	% GC as determined from			Unusual base
	Tm.	Buoyant density	Chem. comp.	
SP8	17.5	84	43	5-hydroxymethyluracil replaces thymine
PBS1	17.5	63	28.2	Uracil replaces thymine
T4	35.9	41.3	34	Hydroxymethylcytosine replaces cytosine. Also glucosylated

† Note that replacement of thymine with uracil does not convert DNA into RNA. The type of nucleic acid, i.e. DNA or RNA, is specified by the sugar moiety.

Base composition heterogeneity

When the DNA from some viruses such as phage λ or the animal adenovirus CELO is centrifuged to equilibrium in CsCl it yields a single band. However, if the DNA molecule is broken in the middle by shearing, the two halves are found to have different buoyant densities and hence different G + C contents. Since mercuric ions preferentially bind to A-T base pairs it is possible to increase the resolution of the two halves by centrifugation in a Cs_2SO_4-$HgSO_4$ gradient. By further shearing the DNA followed by centrifugation in the presence of Hg^{2+} it is possible to separate λ DNA into three components: a 56% GC fragment, a 46% GC fragment and a 41% GC fragment.

Base composition heterogeneity can also be detected from melting profiles. Whereas most DNA's give a smooth melting curve, those which have base composition heterogeneity exhibit a polyphasic curve (Fig. 3.5).

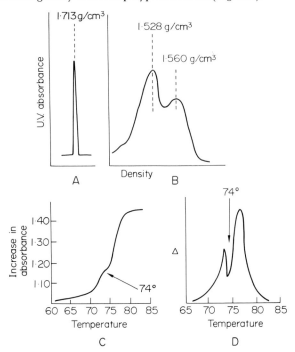

Figure 3.5 Demonstration of base composition heterogeneity in the DNA from chick embryo lethal orphan virus (CELO virus). (A) Equilibrium sedimentation of CELO virus DNA in CsCl showing single sharp peak of absorbance indicating that the DNA is relatively unsheared. (B) Equilibrium sedimentation of sheared DNA in Cs_2SO_4 in presence of Hg^{2+}. Note the presence of two peaks indicating that the DNA fragments do not have the same base composition. (C) Biphasic 'melting' profile of intact CELO virus DNA. (D) Differentiation of the curve shown in C exaggerates the biphasic nature of the melting profile.

Difference in base composition between strands

When certain cellular and viral double-stranded DNA molecules are centrifuged to equilibrium in alkaline CsCl the two strands are observed to have different

46

buoyant densities indicating that they differ in their guanine and/or thymine content. Certain synthetic ribopolymers such as poly I,G preferentially bind to one of the two DNA strands indicating a difference in base composition between the two strands. The resulting differential increase in buoyant density of the two strands is sufficient for the formation of two, usually well resolved, bands in neutral CsCl or Cs_2SO_4 gradients (Fig. 3.6). Viruses whose strands are separable in this way include bacteriophages λ and T7.

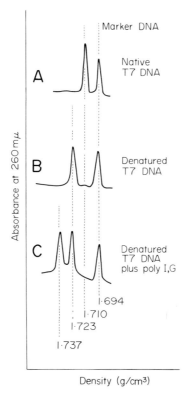

Figure 3.6 Improved separation of the complementary strands of viral DNA with the aid of the ribopolymer poly I,G. (A) Native T7 DNA has a buoyant density of 1.710 g/cc when compared with marker DNA of density 1.694 g/cc. (B) Increase in density of T7 DNA when denatured prior to centrifugation. Note that the DNA sediments as a single species. (C) Resolution of the denatured DNA into two components by hybridization with the ribopolymer poly I,G prior to centrifugation.

Single-stranded ('sticky') ends

The DNA from bacteriophage λ sediments as a single component in sucrose density gradients. Since macromolecules are separated on the basis of size and shape in sucrose gradients it can be concluded that the DNA from λ is homogenous. When the DNA is heated to temperatures below its Tm and slowly cooled, two new components appear, one sedimenting 1.13 times faster and the other 1.41 times faster than native λ DNA. Since the formation of both species increases as

47

the temperature and salt concentration are raised, and since both disappear after melting and quick cooling it is likely that hydrogen bonding is involved. To explain the appearance of the two new components it was proposed that there is a stretch of single-stranded DNA at either end of the molecule and that the single-stranded segments are complementary. The two new components observed would thus consist of hydrogen-bonded circles and dimers (Fig. 3.7) and when

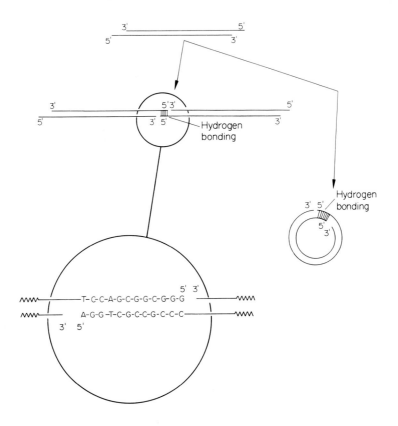

Figure 3.7 Formation of dimers and circles of λ DNA after incubation under conditions favouring annealing. The base composition of the 'sticky ends' is also shown.

such DNA preparations are examined in the electron microscope, circles can be seen. Furthermore, if the DNA is treated with DNA polymerase which has a preference for single-stranded DNA, the product cannot form circles since the single-stranded ends are converted to duplexes. Incubation with exonuclease restores the ability to form circles by regenerating the single-stranded ends.

Terminal redundancy

Genetic studies with bacteriophage T4 suggested that its DNA is terminally redundant; that is, the first few genes at one end of the chromosome also appear at the other end (Fig. 3.8). This was shown biophysically by treating the DNA with exonuclease to produce single-stranded segments followed by incubation under annealing conditions. If the DNA is indeed terminally redundant then 'sticky' ends should have been produced by the exonuclease treatment which should enable the molecules to circularize during the incubation period. Examination of T4 DNA treated in this way revealed the presence of circles thus confirming that it is terminally redundant.

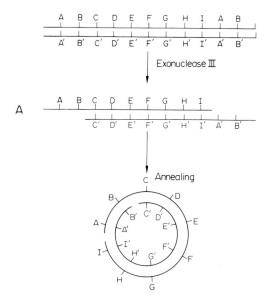

Figure 3.8 Demonstration of terminal redundancy. (A) A terminally redundant molecule is depicted by two parallel lines and the different genes by letters of the alphabet. Following exonuclease treatment, 'sticky ends' are produced which can circularize on incubation under conditions favouring annealing.

Inverted repeats

Treatment of adenovirus DNA with exonuclease does not allow the formation of double-stranded DNA circles. This means that the single-stranded termini exposed by nuclease digestion cannot base pair and that the ends of adenovirus DNA are not terminally repetitious in the conventional manner. However, when a solution of adenovirus DNA is denatured with alkali and then neutralized, both of the DNA strands are capable of forming circles. Since these circles are always of unit length they must be formed by interaction between the 3′ and 5′ termini of the same strand, i.e. there must be an inverted terminal repetition. Two kinds of inverted repeat can be envisaged (Fig. 3.9) and which give rise to different kinds of circular structure. In model A, self-annealed strands would be closed by an 'in-line' duplex segment, whereas strand closure in model B produces a duplex projection or 'panhandle'. In the case of adenovirus DNA, self-annealing produces 'in-line' circles whereas self-annealing of adeno-associated virus DNA produces 'panhandles'.

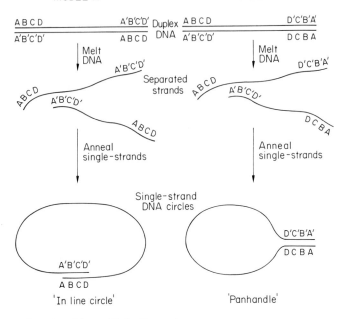

Figure 3.9 Detection of the two kinds of inverted repeat.

Terminal hairpins

The existence of a small proportion of a stable, duplex DNA in the predominantly single-stranded DNA of the parvovirus, minute virus of mice (MVM), has been established by three independent methods. Firstly, a fragment of DNA, about 6.5% of the entire genome, is resistant to the single-strand specific endonuclease S1. Secondly, whereas completely single-stranded DNA elutes from hydroxylapatite columns with 0.12 M phosphate buffer, double-stranded DNA as short as 17 base pairs requires *at least* 0.14 M phosphate buffer to be eluted. Finally, ethidium bromide fluorimetry has been used to obtain a direct, physical measurement of the percentage of double-stranded DNA in several MVM preparations. This method relies on the fact that a pH 12, unstable non-specific base-paired regions are eliminated whereas stable duplex DNA retains its double-stranded structure. When double-stranded DNA is mixed with ethidium bromide an increase in fluorescence is obtained, the intensity of which is proportional to the concentration of double-stranded DNA. Since the increase in fluorescence results from the intercalation of the dye molecule, single-stranded DNA does not interfere with the assay. This method shows that 6% of the genome is double-stranded, in agreement with the endonuclease S1 data.

The fact that exonuclease I, which starts at the 3′ end of single-stranded DNA, digests MVM DNA to the same extent as endonuclease S1 implies that the duplex region is located at the 5′ terminus. Direct proof for this location was obtained by digesting MVM DNA, which was terminally labelled at the 5′ end with $^{32}PO_4$, with endonuclease S1. Since more than 80% of the labelled DNA was not rendered acid soluble by this technique, the 5′ terminus must be in a duplex conformation.

Circular permutation

Once again genetic studies with T4 suggested an unusual feature of its nucleic acid. This virus has a linear DNA molecule and yet its genetic map is circular. An answer to this paradox is to propose that the genes in bacteriophage T4 are

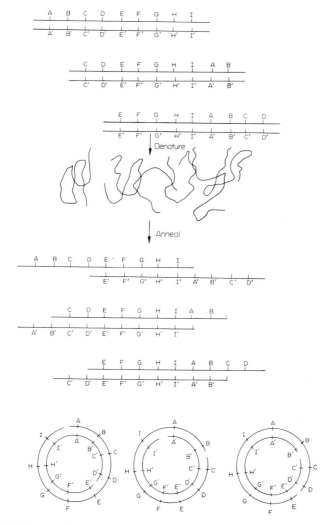

Figure 3.10 Demonstration of circular permutation. A collection of circularly permuted molecules is depicted by parallel lines with the different genes indicated by letters of the alphabet. By denaturing the DNA and then annealing the single strands so produced it is possible to generate molecules with 'sticky ends' which are capable of circularizing.

circularly permuted. If we place all the genes of the bacteriophage round the circumference of a circle and break the circle at any point the result is a collection of linear molecules in which the gene order is unchanged but which have different genes at the ends of the molecules (Fig. 3.10). If such a collection of molecules is denatured and then annealed there is a high probability of complementary strands from different molecules coming together. This will result in the formation of single-stranded ends which are complementary, i.e. 'sticky' ends, which should enable the molecules to circularize. Indeed, when such experiments are done it is possible to observe circular molecules in the electron microscope.

Circular DNA

Not only are some viral DNA molecules capable of circularizing but some actually are circular when extracted from the virus. Such molecules are resistant to exonuclease III, which can only act on those molecules having a free end, but become sensitive after a brief endonuclease treatment. Circular DNA molecules are found with both single-stranded DNA viruses (bacteriophage ϕX174) and double-stranded DNA viruses (simian virus 40). The double-stranded circular molecules can be of two types: a covalently closed double-stranded structure (sometimes called form I, or closed circles) and a double-stranded circular structure in which there is a break in one strand (form II or nicked circles) (Fig. 3.11). There is usually a second difference between these two structures in that the form I isolated form viruses such as SV40 or from ϕX174 infected cells is supercoiled due to a deficiency of turns in the double helix.

Because the shapes of the two molecules are so different they are readily

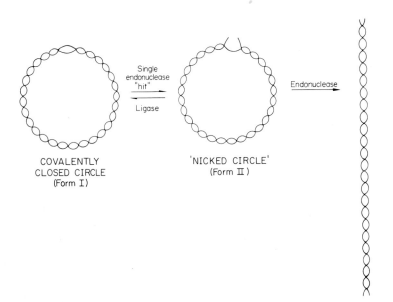

COVALENTLY
CLOSED CIRCLE
(Form I)

'NICKED CIRCLE'
(Form II)

Figure 3.11 Structure of double-stranded circular DNA. The conversion of a nicked circular molecule to the linear form is also shown.

separable in sucrose gradients. It is also possible to separate these circular molecules from linear molecules of the same buoyant density by centrifugation in density gradients containing the drug ethidium bromide. Ethidium bromide is capable of intercalating between the stacked bases of the double helix and as the amount of drug bound increases, the helix untwists until the open form of the circular molecule is produced. Further intercalation introduces excess turns in the double helix resulting in a supercoiling in the opposite sense (Fig. 3.12). Because of the 'opposite' supercoiling there is a limit to the amount of drug which can be bound by a circular molecule but this restraint does not operate with linear molecules. As a consequence of the circular and linear forms binding different amounts of drug their buoyant densities are different and hence they can be separated in CsCl gradients.

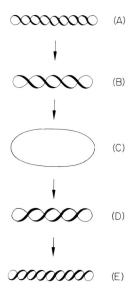

Figure 3.12 Effect of intercalation of ethidium on supercoiling of DNA. As the amount of intercalated drug increases, the double helix untwists with the result that the supercoiling decreases until the open form of the circular molecule is produced. Further intercalation introduces excess turns in the double helix resulting in supercoiling in the opposite sense (note the direction of coiling at B and D).

Protein linkers

Examination in the electron microscope of partially disrupted adenoviruses reveals circular DNA, sometimes in the form of supercoils, whereas DNA extracted from adenoviruses by treatment with proteolytic enzymes, detergent and phenol consists of linear duplex molecules. Adenovirus DNA is not permuted and lacks sticky ends and terminal repetitions, structural peculiarities found in other viral DNA molecules and which enable them to form circles. So far, the only structural peculiarity detected in adenovirus DNA is the presence of an inverted terminal repetition (see Fig. 3.13) but this does not permit circularization of duplexes. The explanation for the circular DNA molecules found in partially

53

disrupted virions, but not in extracted DNA, is the presence of a protein linker which joins the two ends of the duplex. This protein linker is destroyed by the proteolytic enzymes and detergent used in extracting the DNA. When the virus is disrupted by guanidinium chloride and deproteinized with chloroform and isoamyl alcohol the protein linker can be kept intact. DNA isolated in this way sediments in sucrose gradients faster than the linear duplexes obtained by conventional extraction methods (Fig. 3.13A) and electron microscopy reveals that up to 90% of the DNA is in the form of circles and oligomers. Treatment of these circles and oligomers with detergent or proteolytic enzymes converted them to linear duplex monomers (Fig. 3.13B).

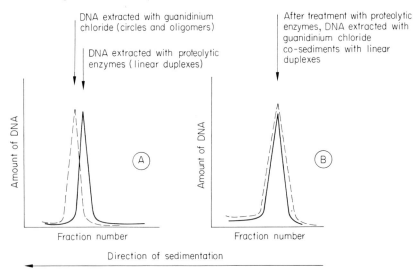

Figure 3.13 Sucrose gradient sedimentation of DNA extracted from adenoviruses by two different methods. (A) Separation of circles and oligomers from linear duplexes. (B) Conversion of circles and oligomers to linear duplexes by treatment with proteolytic enzymes.

Circular RNA

The RNAs of the causative agents of potato spindle tuber (PSTV) and citrus exocortis disease (CEV) exhibit hyperchromicity upon heating indicating the presence of base-pairing. Detailed analysis of the thermal denaturation profiles suggests that there are relatively long double helical regions rather than short hairpin sections as in tRNA. When examined in the electron microscope, native RNA from PSTV and CEV can be seen as rod-shaped or dumb-bell-shaped molecules. Denatured RNA from PSTV and CEV, on the other hand, exhibits a variety of structures ranging from rod-shaped molecules (renaturation during preparation for electron microscopy?) to single-stranded circles. All attempts to label the 5' and 3' ends, using the techniques outlined for reovirus (p. 56), have been unsuccessful suggesting that there are no free ends. These data suggest that the RNAs from PSTV and CEV exist as covalently-closed circular molecules with considerable intra-molecular base-pairing. It is worth noting

54

that with molecular weights of approximately 120 000 daltons these are by far the smallest circular nucleic acids known.

Terminal caps

Messenger RNAs are synthesized, translated and possibly also degraded in a $5' \rightarrow 3'$ direction. Consequently their $5'$ termini are of some interest particularly with regard to their ability to regulate genetic expression (see Chapter 8). In prokaryotes, and viruses of prokaryotes, the $5'$ end of many mRNAs is a triphosphorylated purine corresponding to the residue that initiated transcription. By contrast, most eukaryotic cellular and viral mRNAs as well as native nucleic acid from some RNA viruses of eukaryotes, have been found to be modified at the $5'$ end. The modification consists of a 'cap' that protects the RNA at its $5'$ terminus from attack by phosphatases and other nucleases and promotes mRNA function at the level of initiation of translation (see also p. 109).

The general structural features of the $5'$ cap are shown in Fig. 3.14. The terminal 7-methylguanine and the penultimate nucleotide are joined by their

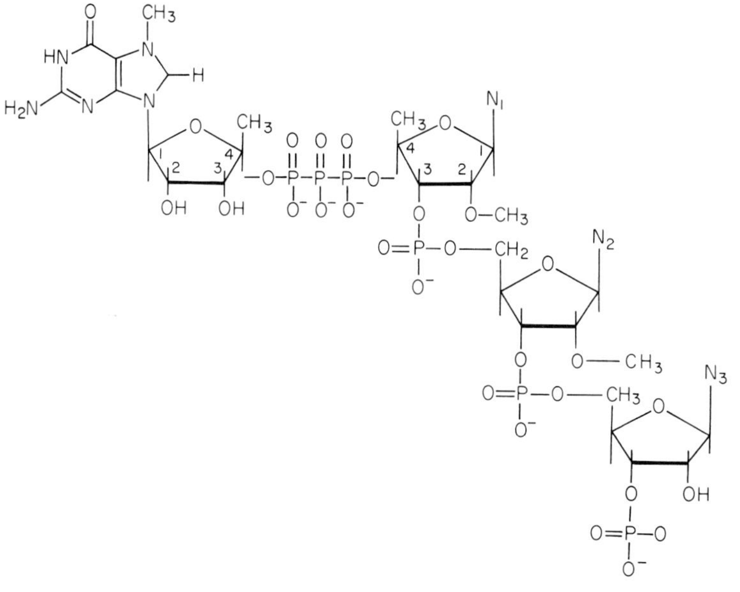

Figure 3.14 Structure of a capped RNA molecule. Note especially the $5'$-$5'$ phosphodiester linkage. N_1, N_2 and N_3 can be any of the four nucleic acid bases: The $2'$ O-methyl group on the ribose of N_1 nucleotides is always present but the methyl group on N_2 is present in some cases only.

$5'$-hydroxyl groups through a triphosphate bridge. This $5'$-$5'$ linkage is inverted, relative to the normal $3'$-$5'$ phosphodiester bonds in the remainder of the poly-nucleotide chain. As a consequence of the inversion of the terminal 7-methyl-guanine residue the capped end of the RNA has a $2'3$-cis diol that provides the basis for several chemical manipulations of caps. For example, since free $2'3'$-hydroxyls can form complexes with borate ions, caps or capped oligonucleotides can be purified by affinity chromatography on substances such as dihydroxyboryl cellulose. This cis diol is oxidized by periodate treatment to the dialdehyde which can then be reduced with ^{3}H-borohydride to label the terminus. Alternatively, exposure of the oxidized mRNA to aniline causes β-elimination of the 7-methyl-guanine dialdehyde, yielding a 5′-triphosphate-ended RNA molecule.

Segmented genomes

Chemical analysis of the reovirus genome showed it to have a mol. wt. greater than 10^7 daltons, but the observed sedimentation rate of viral RNA was lower than that expected for a molecule of this size. Further studies showed that regardless of the amount of care exercised in extracting the viral RNA, a trimodal distribution of sizes was always found by velocity sedimentation of the RNA on sucrose gradients. When subjected to electrophoresis, each size class could be further separated resulting in the recognition of 10 different molecules of RNA. These 10 fragments showed no base sequence homology in hybridization tests and so could not have arisen by random fragmentation of the genome.

The sum total of the mol. wts. of the 10 fragments corresponds to the size of the viral genome as estimated by chemical means suggesting that the intact virus contains one copy of each fragment. Consequently, does the genome exist in the intact particle as a single molecule, which is susceptible to breakage at fixed points during extraction, or as a collection of 10 different fragments? The reovirus genome consists of double-stranded RNA and we can propose four models for its structure (Fig. 3.15). In trying to decide between these models, a technique has been developed for specifically labelling the 3′ ends of the RNA, regardless of whether the RNA has been extracted or is still in the intact virion.

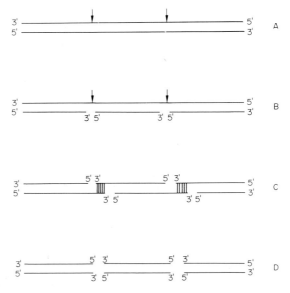

Figure 3.15 Schematic models for the double-stranded RNA genome of reovirus. The arrows represent hypothetical weak points in the viral genome.

The technique involves oxidizing the RNA with periodate resulting in the formation of a dialdehyde at the free 2′, 3′-OH positions. This dialdehyde is then reduced with tritiated borohydride which introduces four tritium atoms per 3′ terminal (Fig. 3.16). In the actual experiments reovirus was labelled with ^{32}P to provide an RNA marker, purified, and the virus preparation divided into two. RNA was extracted from one part, oxidized with periodate, and reduced with tritiated borohydride. The other part of the virus was oxidized and the

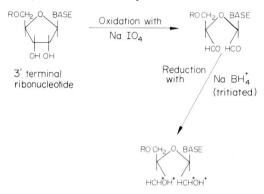

Figure 3.16 Tritium labelling of 3′ terminal ribonucleotide. The terminal ribonucleotide is oxidized to a dialdehyde and this is then reduced with tritiated borohydride of high specific activity. Tritiated atoms are indicated with asterisks.

RNA was extracted and then analysed by gel electrophoresis. Results showed that all ten fragments were labelled, even when the RNA had not been extracted prior to oxidation and so we can eliminate model A immediately. According to model B, one strand is continuous and the other is nicked in nine places. Oxidation of this structure *within the virion* will modify only the left-hand segment at both 3′ terminals, while the other nine segments will be modified at one 3′ terminus only. This means that during the reductive step, nine of the ten sub-units will incorporate half as much tritium as when oxidation is done on extracted double-stranded RNA. In the actual experiments, quantitative analysis of the tritium distribution showed that both samples of double-stranded RNA incorporated tritium to about the same extent in all classes and we can thereby eliminate model B.

If model C were correct then one would expect to find short single-stranded tails on the RNA fragments. When RNA labelled with ^{32}P is extracted from the virion and treated with nucleases which attack single-stranded RNA at either the 3′ or 5′ end, no release of ^{32}P was observed. We can thus rule out model C suggesting that model D represents the structure of the reovirus genome.

Other viruses which probably have segmented genomes include cytoplasmic polyhedrosis viruses which also have double-stranded RNA and influenza, Rous sarcoma virus and brome mosaic virus which have single-stranded RNA. Brome mosaic virus differs from the others since not all the segments are contained in the one particle. Instead, the four observed classes (1-4) of RNA are contained in three different particles (Fig. 3.17) and experiments with isolated RNA show that a mixture of the three largest classes is required for infectivity. The smallest fragment, No. 4, is dispensable and pancreatic ribonuclease digests show that it contains nucleotide sequences also found in component 3. As yet, the significance and origin of this fragment are unknown. Details of other multicomponent viruses are given in Chapter 18.

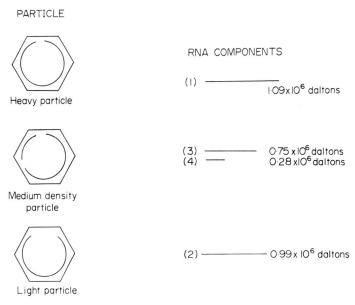

Figure 3.17 Relationships between the RNA and nucleoprotein components of brome mosaic virus. The RNA components are numbered from 1 to 4 on the basis of size.

HERPESVIRUS DNA HAS A NUMBER OF STRUCTURAL PECULIARITIES

DNA from herpes simplex virus type I (HSV1) is terminally redundant since incubation of exonuclease-treated DNA under annealing conditions results in the formation of circles (Fig. 3.8). However, HSV1 DNA is considerably more complex than this. When the isolated *single*-strands are self-annealed, two kinds of molecules of particular interest can be seen by electron microscopy. The first of these, designated type A, are ones in which one terminus annealed to an internal region to form a structure consisting of a small single-stranded loop, a short double-stranded region, and a long single-stranded tail (Fig. 3.18). Molecules designated type B are those in which both termini annealed to internal inverted sequences to yield a structure consisting of a large and a small single-stranded loop bridged by a double-stranded region. These results indicate that the two terminal sequences ABCDE and HGFAB are also found internally in the form of adjacent inverted repeats (Fig. 3.18) and that the unique DNA segment, called L, separating ABCDE from E′D′C′B′A′ is longer than the unique segment called S, which separates B′A′F′G′H′ from HGFAB. Although the two terminal sequences are not identical, the sequence A-B must be present in both for the molecule is known to be terminally redundant.

58

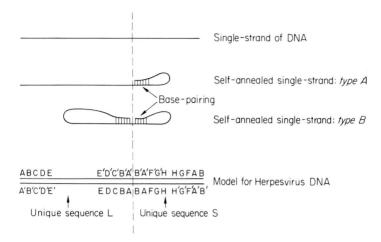

Figure 3.18 Detection of inverted repeats in herpesvirus DNA and the proposed model for the molecular structure of the DNA. Note that the sequence AB appears at both ends of the molecule (terminal redundancy).

The endonuclease H*ind* III (see Chapter 6) is site specific and only cleaves DNA at sites containing the sequence

Whereas HSV1 and DNA has a molecular weight of 97×10^6 daltons, treatment with endonuclease H*ind* III produces 14 fragments whose combined molecular weight is 160×10^6 daltons. Measurement of the relative molar concentrations of the 14 fragments suggests that six 'major' fragments, accounting for 60% of the genome, occur once in every molecule, four 'minor' fragments occur once in every two molecules and the remaining four fragments occur once in

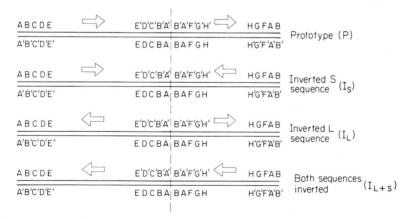

Figure 3.19 The four possible genome arrangements which can result from inversions of the long and short unique regions of HSV1 DNA.

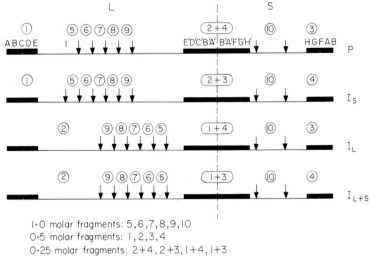

1·0 molar fragments: 5,6,7,8,9,10
0·5 molar fragments: 1,2,3,4
0·25 molar fragments: 2+4,2+3,1+4,1+3

Figure 3.20 Production of DNA fragments in different molar amounts by endonuclease cleavage of the four forms of HSV1 DNA. Terminal and internal repeats are indicated by solid bars. The arrows indicate the sites of endonuclease cleavage and the different fragments are identified by numbers.

every four molecules. These results are easily explained if HSV1 DNA exists in 4 forms (Fig. 3.19) depending on the orientation of the L and S segments. The way in which the endonuclease generates 14 fragments present in differing molar amounts can be seen by reference to Fig. 3.20.

One important consequence of these novel structural features relates to the genetics of herpesviruses. Although mutations within the L and S segments map in a linear fashion, all crosses between markers in L and S would be expected to give maximal recombination frequencies and therefore the genetic map for the entire herpesvirus genome should consist of two separate linkage groups.

OVERLAPPING GENES

In recent years the genomes of a number of viruses have been completely sequenced and the list of those sequenced grows monthly. The first complete sequence of a viral RNA was that of bacteriophage MS2 which was achieved by Fiers' group in 1976. In 1977 Sanger and his collaborators published the complete sequence of the genome of the single-stranded DNA bacteriophage ϕX174. The elegant techniques used in sequencing genomes are too complex to be considered here but the interested reader should consult the original papers cited on page 240.

Once the complete sequence of a genome is known it is possible to scrutinize it for information on its overall genetic organization. For example it is possible to locate the start signals for each gene since the initial amino acid in all protein chains is methionine which is specified by a single codon, ATG. Not all methionine residues are at the start of a protein so not all ATG sequences will be signals for initiating a protein. Rather, the ATG codon must be preceded by a sequence which tells the ribosome to begin translation. Although the exact sequence of these ribosome recognition signals varies they are similar enough to be easily recognizable. Once the start of a gene is located it is possible to locate its terminus by moving along the sequence until one of the three termination codons TGA, TAA or TAG in the correct reading frame is encountered.

Detailed studies of a number of bacterial operons suggested that coding regions of DNA were clearly separated by non-coding regions. This idea was strengthened when sequencing of the complete MS2 genome showed that the 3 genes were separated by two non-coding regions, one 26 nucleotides long, the other 36 nucleotides long. Examination of the complete sequence of the ϕX174 genome suggests that there is no absolute requirement for non-coding spacer regions between genes. Firstly, in three instances the termination codon of one region overlaps the initiation codon of the next gene leaving no space for an untranslated region (Fig. 3.21). Thus the end of gene A overlaps the start of gene C, the end of gene C overlaps the start of gene D and the end of gene D overlaps the start of gene J. Secondly, and much more surprising, two small genes are entirely contained within two larger genes (Fig. 3.21). Thus gene B is contained

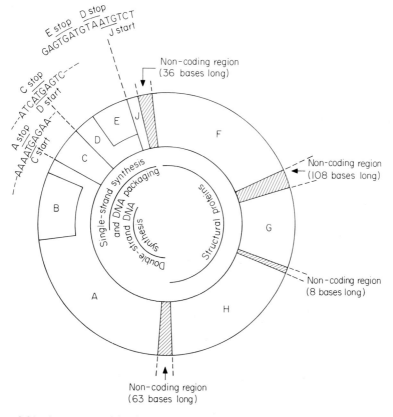

Figure 3.21 Arrangement of the nine genes of bacteriophage ϕX174 showing the DNA sequences in the vicinity of the overlaps between genes A,C,D, and J. Note that genes B and E are entirely located within genes A and D respectively. Note also that genes with similar function are located adjacent to each other.

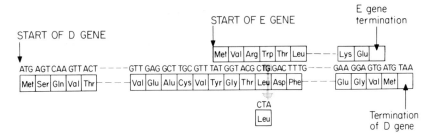

Figure 3.22 DNA sequence at the start and finish of the overlapping genes D and E showing how the two genes are translated in different reading frames. The shaded guanine residue is the base which is converted to an adenine residue in the E gene amber mutant described in the text.

within gene A and gene E within gene D. In both cases, however, the smaller genes are encoded in different reading frames from the larger genes (Fig. 3.22).

How were overlapping genes discovered? Conventional genetic mapping of φX174 had located gene E between genes D and J but correlation of the amino acid sequence of D and J proteins with the nucleotide sequence clearly indicated that genes D and J were contiguous. This apparent paradox was resolved by DNA sequence analysis of an amber mutant of gene E. In the mutant DNA a guanine residue was changed into an adenine residue and this change was in a region of the DNA that codes for the D protein. In a particular reading frame the effect of this change was to convert a tryptophan codon, TGG, into the termination codon TAG, thus explaining the amber phenotype (premature chain termination) of the mutant. The reason that the D gene product is unaffected in the E gene mutant is simple. The reading frame for the D gene is different (Fig. 3.22) and the change from guanine to adenine converts the leucine codon CTG into CTA which also codes for leucine.

THE IMPORTANCE OF STRUCTURAL PECULIARITIES

The study of the structural peculiarities of viral nucleic acids is important for several reasons. Firstly, these structural peculiarities, particularly the presence of unusual bases, may help the virus to subvert the cell's biosynthetic machinery and redirect it to the production of new virus. Secondly, any models which may be proposed for the replication of these nucleic acids must take into account their structural peculiarities. In this context it is worth pointing out that many viral nucleic acids are either circular or capable of circularizing and thus resemble the chromosomes of *E. coli, Streptomyces* and the mycoplasma. We shall return to this theme in Chapter 6 when we discuss the 'rolling circle' model for DNA replication. Thirdly, these structural peculiarities frequently have important practical implications. For example, it is highly unlikely that both strands of a DNA molecule are transcribed *in toto* into messenger RNA. The ability to separate the two DNA strands permits us to determine which regions of the molecule are transcribed and to assess the importance of the results with regard to regulation of development. Again we shall return to this theme later (Chapter 8).

4 The process of infection: I adsorption and penetration

The process of infection begins with the coming together of a virus particle and a susceptible host cell but this union comes about by different means with each of the three types of virus; viz. plant viruses, animal viruses and bacteriophages. With bacteria in liquid culture, the interaction with phage most likely occurs by simple diffusion since particles the size of bacteria and viruses are in constant Brownian motion when suspended in liquid. If phage are mixed with susceptible bacteria and the mixture centrifuged at time intervals thereafter, an unadsorbed phage will remain in the supernatant. In this way it is possible to measure the rate of adsorption of phage to bacteria and experiments of this kind have led to the following relationship:

$$\frac{dP}{dt} \, \alpha \, P, \, B \text{ or } \frac{dP}{dt} = KPB$$

where dP/dt is the rate of adsorption and P and B the free phage and bacterial concentrations respectively. The constant of proportionality, K, is called the adsorption rate constant. From the equation it is clear that there is no active participation by either the phage or the bacterium. It should be noted that the relationship explains the theoretical basis for using dilution as a means of reducing or preventing phage adsorption, for a tenfold dilution will decrease the adsorption rate one hundredfold.

Diffusion of animal viruses is probably the force influencing their union with animal cells in tissue culture since their attachment is independent of temperature except insofar as this affects Brownian motion. This may not be true with animal viruses and intact animals but it is difficult to design suitable experiments with the latter! In most cases plants become infected with viruses following mechanical damage to the plant, very often as a result of the activities of virus-carrying insects. Consequently, the way in which the union of virus and cell occurs is not so important in plant systems. However, when viruses are transmitted in plants as a result of grafting, diffusion through the vascular system is most likely responsible.

ADSORPTION

Experiments on adsorption of phage to bacteria, like that detailed above, really only measure phage particles that are *irreversibly* bound to bacteria. Phage particles can also be *reversibly* adsorbed and this can be detected by adsorbing phage to cells in suspension and centrifuging the complexes into a pellet. When the pellet is resuspended in fresh medium and the phage-bacterium complexes are again centrifuged down, many infective phages are left in the supernatant.

These infective phages, initially reversibly bound to the cells, were released from the cells when the fresh medium was added.

It has been postulated that reversible adsorption, due to electrostatic forces, initially binds the phage particle to the cell so that processes making the bond a permanent one can take place. With some phages adsorption may be favoured or inhibited by varying the concentration of certain ions in the medium, in particular Mg^{2+} and Ca^{2+}. Eventually the phages become irreversibly adsorbed so that their removal with retention of activity becomes impossible and thus there is some doubt whether the reversible stage is necessary to virus adsorption.

ADSORPTION OF BACTERIOPHAGES TO THE CELL WALL

Most bacteriophages adsorb to the cell wall but this is not the only adsorption site since other phages can adsorb to the pili, flagella or capsule of the host. Most of the tailed bacteriophages adsorb to the cell wall and do so by the tip of their tail. With those phages whose head is much larger than the tail, diffusion will tend to cause the tail to oscillate more than the head making it more likely that the tail will collide first with the cell. The chemical nature of the adsorption site on the cell wall has been elucidated for some *Salmonella* phages and has proved most instructive. Figure 4.1 shows the structure of the O-antigen of wild-type *Salmonella typhimurium* and certain cell wall mutants derived from it. When the mutants are tested for sensitivity to different phages it becomes apparent that each has a characteristic sensitivity pattern. For example, the presence of the O-specific side chain makes a cell resistant to phages 6SR, C21, Br60 and Br2 but sensitive to P22 and Felix O. The absence of the O-specific side chain, as in *rfb* T mutants now makes the cell sensitive to 6SR, Br60 and Br2 and resistant to P22 and Felix O. Thus, if the lipopolysaccharide is the adsorption site for a particular phage it is possible to determine precisely the composition and structure of the receptor by testing the different mutants for sensitivity to that phage. Experiments of this type are not only of interest to virologists but also to cell wall chemists. For example, suppose we wished to isolate an *rfc* mutant of *Salmonella*. In the absence of a selective procedure this would be a tedious task. However, if a bacterial lawn showing plaques of P22 is incubated for several days, small colonies arise within the plaques which represent the growth of phage-resistant mutants. It is clear that these resistants could belong to any one of the mutant classes shown in Fig. 4.1. However, by simultaneously selecting for resistance to 6SR, C21, Br60 and Br2 we can eliminate all but the *rfc* class which should be sensitive to Felix O.

The best documented example of phage adsorption is that of phages T2 and T4. As will be recalled from Chapter 2, these viruses have a complex structure including a tail, baseplate, pins and tail fibres. The initial attachment of these phages to the receptors on the bacterial surface is made by the distal ends of the long tail fibres (Fig. 4.2). The long tail fibres which make the first attachment, bend at their centre and their distal tips contact the cell wall only some distance from the midpoint of the phage particle. After attachment, the phage particle is apparently brought closer to the cell surface. When the base plate of the phage is about 100 Å from the cell wall, contact is made between the short pins extending from the base plate and the cell wall, but there is no evidence that the base plate itself is attached to the cell wall.

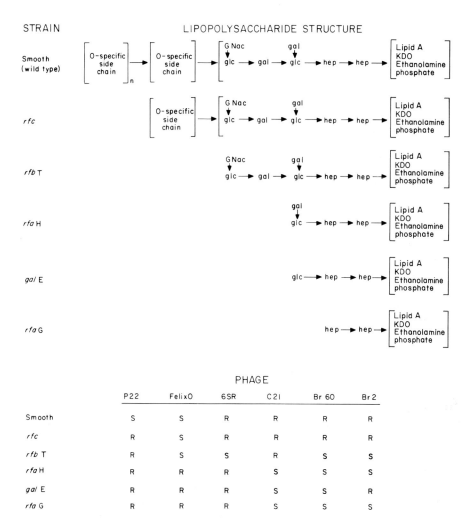

Figure 4.1 Correlation of phage sensitivity with lipopolysaccharide structure of the cell wall. The upper part of the diagram shows the structure of the lipopolysaccharide in different mutants of *Salmonella typhimurium*. Abbreviations: Glc, glucose; Gal, galactose; GNac, N-acetylglucosamine; Hep, heptose; KDO, 2-keto-3-deoxyoctonic acid. The table shows the sensitivity of the different mutants to several bacteriophages. Abbreviations: S, sensitive; R, resistant.

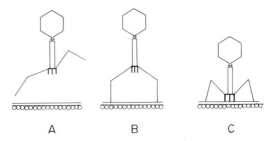

A   B   C

Figure 4.2 Schematic illustration of the major steps in the attachment of bacteriophage T4 to the cell wall of *E. coli*. (A) Unattached phage showing tail fibres and tail pins (cf. Fig. 2.9). (B) Attachment of the long tail fibres. (C) The phage particle has moved closer to the cell wall and the tail pins are in contact with the wall.

Other adsorption sites for bacteriophages

Not all tailed phages adsorb to the cell wall. Some such as phage Χ(*chi*) and PBS1 attach to the flagella. The tip of the tail of these phages has a kinky fibre on it which wraps round the filament of the flagellum and the phage then 'slides' down the filament till it reaches the base of the flagellum (Fig. 4.3). Other tailed phages adsorb to the capsule of the cell.

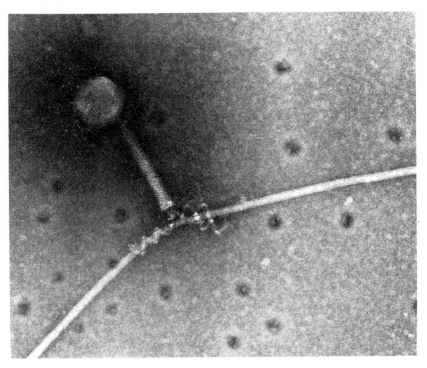

Figure 4.3 Attachment of bacteriophage Chi to the filament of bacteria flagella (courtesy of Dr J. Adler).

The other important adsorption organs are the sex pili. Bacteria which harbour the sex factor (F) or certain colicins or drug resistance factors produce pili and two classes of phage have been shown to adsorb to these pili. The filamentous single-stranded DNA phages adsorb to the tips of the pili while the spherical RNA phages adsorb along the sides of the pili (Fig. 4.4). These phages are particularly useful to microbial geneticists because they offer a ready means of establishing whether cells harbour pili of these types.

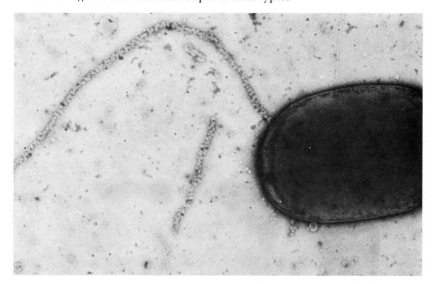

Figure 4.4 Adsorption of spherical RNA phages to the sex pili of *Escherichia coli* (courtesy of Dr C. C. Brinton).

THE ADSORPTION OF ANIMAL VIRUSES

With animal cells in tissue culture, it has been possible to measure attachment of animal viruses and to determine the effects of temperature and the ionic environment. Just as with phages, attachment appears to be electrostatic and is considerably affected by the ionic environment. However, little information is available about cellular receptors for animal viruses except for poliovirus and influenza virus.

Poliovirus attaches efficiently to cultured human and monkey cells whereas insusceptible cells fail to adsorb detectable amounts. When susceptible cells are disrupted by freezing and thawing they release a virus binding receptor substance to which poliovirus will attach and these isolated receptors have the same co-factor requirements as intact cells. The cellular receptor is destroyed by treating the cells with proteases and is presumably a protein. Receptors can be saturated by adding an excess of virus particles. Calculations show that the saturation value for poliovirus is about 3×10^3 particles per cell. This reflects a limitation of receptors since this number of poliovirus particles would occupy around 0.3% of the total cell surface area. Insusceptible cells lack the virus receptor hence the poliovirus cannot attach. The cells are otherwise perfectly capable of supporting virus multiplication as can be shown by artificially persuading the cells to take up poliovirus RNA. Infectious particles are produced but of course these cannot reinfect the receptor-less cells.

When influenza virus is mixed with red blood cells the cells agglutinate (*haemagglutination*), and this can be used as a means of titrating the virus simply by determining what dilution of the virus no longer causes haemagglutination. If cells which have been agglutinated at 4°C are warmed up to 37°C, the cells separate and the virus eluted can again cause agglutination if supplied with fresh cells. However, the cells from which the virus was eluted can no longer agglutinate if mixed with fresh virus and when first discovered the activity of an enzyme was invoked to explain this behaviour. Later an enzyme called *receptor destroying enzyme* (RDE) was isolated from *Vibrio cholera* which could render erythrocytes inagglutinable. It was soon shown that influenza virus adsorbs to cells lining the respiratory tract in much the same way since RDE treatment of the respiratory tract makes the animal insusceptible to influenza infection.

RDE is a highly specific enzyme which hydrolyses acetylated neuraminic acid when it is in a terminal position in an oligosaccharide. Hence RDE is a neuraminidase. Influenza viruses therefore have terminal acetylated neuraminic

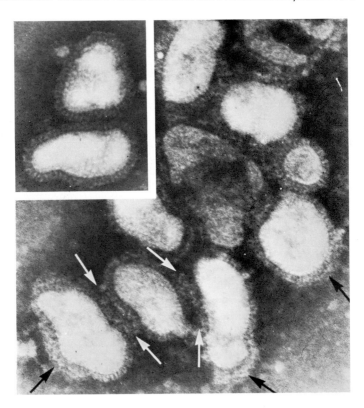

Figure 4.5 Electron micrograph of an influenza virus incubated with specific antibody (arrowed) which is seen as fuzzy material attached to the particles. Spikes are shown clearly in virus without antibody (inset). (Courtesy of R. W. Compans.)

acid as their receptor. These molecules are ubiquitous on the surface of animal cells and are usually part of an oligosaccharide covalently bound to a protein which is itself inserted into the plasma membrane.

When examined in the electron microscope, negatively-stained preparations of influenza virus reveal the presence of an envelope covered with spikes (Figs. 4.5; 2.10). The spikes, when isolated, have been shown to be of two morphological types, one of them possessing neuraminidase activity and the other being the haemagglutinin. The infectivity of influenza virus in cultured cells can be completely neutralized by antibody to the haemagglutinin but is unaffected by antibody directed specifically against the viral neuraminidase. Thus, only the haemagglutinin sub-units are important in attachment and entry into cells in culture. However, the neuraminidase is important in natural infections of the respiratory tract since it frees virus which attaches to acetylated neuraminic acid residues on mucus and allows the virus another chance of attaching to a cell receptor. When this occurs the adsorbed virus is normally internalized by the cell before the viral neuraminidase can release it.

PENETRATION

Penetration by bacteriophage T2

The experiments of Hershey and Chase outlined in Chapter 1 indicated that only the nucleic acid entered the cell during bacteriophage T2 infection. The way in which this occurs has now been elucidated and is a complex but fascinating story which we can only sketch briefly here. The tail of the bacteriophage is contractile and in the extended form consists of 24 rings of sub-units surrounding a core. Each ring consists of six sub-units of one size or six sub-units of a larger size. Following adsorption the tail contracts resulting in a merging of the small and large sub-units to give 12 rings of 12 sub-units. The tail core, which is not contractile, is pushed through the outer layers of the bacterium with a twisting motion and contraction of the head results in the injection of the DNA into the cell. This process is probably aided by the action of the lysozyme which is built into the phage tail. There are 144 molecules of ATP built into the sheath and the energy for contraction most likely comes from their conversion to ADP. The phage has been likened to a hypodermic syringe, and the various steps in penetration are shown in Fig. 4.6.

Penetration by RNA bacteriophages

Many bacteriophages do not possess contractile sheaths and so the way in which their nucleic acid enters the cell is not known. Hershey-Chase type experiments with the filamentous DNA phages, which attach to the sex pili, suggest that both the DNA and the coat protein enters the cell and it has been postulated that after adsorption the phage-pilus complex is retracted by the host cell. However, when similar experiments are performed on the RNA phages, which also attach to the sex pili, the results obtained are similar to those for T2, i.e. the phage coat protein does not enter the cell. Indeed, following adsorption the RNA phages are rapidly eluted again. Examination of the eluted phage in sucrose gradients shows that 70% of them sediment as intact 78S particles, while the remaining 30% lack RNA and sediment as 42S particles. Treatment of the eluted 78S particles with RNase converts them to 42S particles whereas phage which have not been allowed to interact with pili remain RNase resistant. An analogous sequence of events occurs when poliovirus interacts with cells carrying the appropriate receptors except that the RNA of the particles is still RNase resistant.

Since the A protein of the single-stranded RNA phages is involved in adsorption it is likely that the RNase sensitivity of the eluted 78S particles stems from loss of this protein. It is, in fact, possible to deduce the course taken by the A protein by locating the radioactivity ^{3}H-histidine labelled phage following its interaction with host cells. Such experiments are based on the fact that the coat protein of the RNA phages lacks the amino acid histidine whereas the A protein

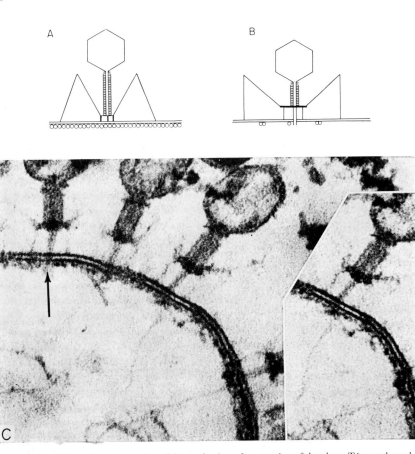

Figure 4.6 Schematic representation of the mechanism of penetration of the phage T4 core through the bacterial cell wall. (A) The phage tail pins are in contact with the cell wall and the sheath is extended. (B) The tail sheath has contracted and the phage core has penetrated the cell wall; phage lysozyme has digested away the cell beneath the phage. (C) Electron micrograph of T4 adsorbed to an *E. coli* cell wall as seen in thin section. The needle of one of the phages can be seen to penetrate just through the cell wall (arrow). The thin fibrils extending on the inner side of the cell wall from the distal tips of the needles are probably DNA.

70

contains approximately 4.5 residues of this amino acid per molecule. When ^{3}H-histidine-labelled phage are allowed to adsorb and elute from pili, the histidine label is completely absent from the 42S particles and only a small amount is present in the 78S particles. It can thus be concluded that the A protein is released from the phage particles following adsorption.

Further experiments on the RNA phages have shown that the viral RNA and the A protein are injected into the cell in approximately equimolar amounts, and that their kinetics of penetration are similar. This was shown by infecting cells with phage labelled with [32]P (RNA) and ^{3}H-histidine (A protein). After adsorption of the phage has taken place, the cells are depilated by forcing them through a narrow-gauge hypodermic needle, and freed from the pili by repeated centrifugation and resuspension in fresh medium. The ^{32}P and ^{3}H radioactivity remaining with the cells was determined and after conversion to phage equivalents it was clear that similar amounts of RNA and A protein were taken up by the cells.

Penetration by animal viruses

Animal cells do not have a rigid cell wall but are bounded by a plasma membrane which is a very mobile and active structure (Fig. 4.7A). Thus cells are constantly taking samples of their immediate environment by pinocytosis or carrying out the reverse process to export from the cell substances such as enzymes, hormones or neurotransmitters (Fig. 4.7B).

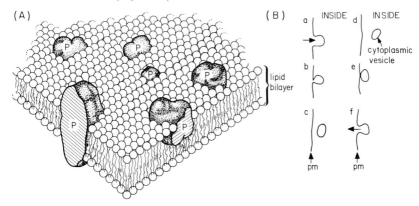

Figure 4.7 (A) 'Lipid sea' model of plasma membrane structure devised by Singer and Nicholson. Proteins (P) may span the lipid bilayer and are free to move laterally like icebergs. (B) Pinocytosis by a plasma membrane (pm) inwards (a,b,c) and outwards (d,e,f). ((A) redrawn from Singer S. J. & Nicholson G. L. (1972) in *Science* **175**, p. 723.)

Viruses can only infect a cell if it has an appropriate receptor molecule on its outer surface and different receptors are required by different viruses. However, the precise details of how viruses enter animal cells after attaching to a cell receptor are not clear. One of the major difficulties in this study is that the majority of virus particles do not successfully initiate multiplication. These 'non-infecting' particles often outnumber the infecting particles by 1000:1. Neither electron microscopy nor biochemical studies can distinguish between non-infecting particles and infecting particles which both contain the viral genome. However, it is clear that enveloped viruses can penetrate into cells by fusing with the plasma membrane while both enveloped and non-enveloped viruses can penetrate by pinocytosis. Fusion (Fig. 4.8A) results in the release of the viral genome into the cytoplasm but after pinocytosis (Fig. 4.8B) the virus is contained in a vesicle of plasma membrane. Various possibilities are open to release the virus from its vesicle. For instance, it may attract the attention of lysosomes which have enzymes capable of digesting away the virus coat. However, lysosomes also contain nucleases which would destroy the viral nucleic acid. A more likely alternative (illustrated in Fig. 4.8B) is that the vesicle fuses with an intracellular membrane. The virus coat is then presumably modified in some unknown way, so that the nucleic acid is released.

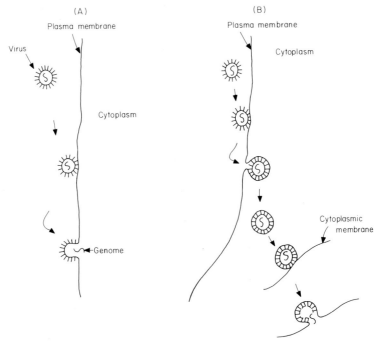

Figure 4.8 Penetration and uncoating of viruses in animal cells. (A) By fusion of viral and plasma membranes, and (B) by pinocytosis followed by fusion with an internal cytoplasmic membrane. The latter part is speculative.

It should be pointed out that penetration and uncoating do not end with naked nucleic acid free in the infected cell. The nucleic acid may remain associated with viral internal proteins which could include enzymes necessary for nucleic acid synthesis, or if the genome also serves as message, will associate with cellular structures such as ribosomes. Equally the nucleic acid will not be stretched out (as illustrators frequently draw it!) but will 'tangle' up with itself and associated proteins into the most energetically favourable state.

One of the striking aspects of penetration of animal cells by viruses is its inefficiency. In poliovirus infections the majority of the RNA of the infecting virus is degraded as a result of interaction with cells, presumably because only a few of the many viral receptors lead on to successful infection. Thus we can account for the very high physical particle: infectious particle ratio mentioned earlier.

An unusual situation is the uncoating of poxviruses. This takes place in two stages as shown by the following experiment. Cells were infected with rabbit pox virus which was labelled in both the nucleic acid and phospholipid components and the susceptibility of the labelled DNA to the action of DNase was used as an index of uncoating. When cells were disrupted at intervals following infection it was observed that whereas phospholipid is released without appreciable delay, release of the DNA, in a form that was susceptible to DNase, did not occur for 1-2 hours. The lag in appearance of DNase sensitivity could be abolished by infecting cells with unlabelled virus before adding labelled virus. However, the lag could be prolonged indefinitely by the addition of puromycin and fluoro-phenylalanine (inhibitors of protein synthesis) or actinomycin D (inhibitor of RNA synthesis). An explanation for these results is provided by the isolation from highly purified virions of a DNA dependent RNA polymerase activity which can transcribe viral DNA. Release of the DNA in a form which is susceptible to DNase probably depends on the activity of an enzyme which is transcribed from viral DNA early in infection by the virus associated polymerase. Release of phospholipid, on the other hand, is mediated by a pre-existing cellular enzyme.

Penetration of plants by viruses

The viral infection of plants occurs by quite different means to those we have discussed for bacteria and animals. Most plants have rigid cell walls of cellulose and consequently viruses must be introduced into the host cytoplasm by some traumatic process. This is why the application of carborundum to leaves increases the number of lesions produced during a local lesion assay of virus-containing material. The majority of plants that become infected naturally do so because virus-carrying animals feed upon them. However, this transmission is not a casual process which occurs whenever any animal chances to feed on an uninfected host plant after feeding on an infected one. Rather, the transmission of most plant viruses is a highly specific process requiring the participation of particular animals as vectors. Some viruses such as TMV require no vector and can be transmitted mechanically and these are the only viruses which can be assayed by the local lesion method. Most plant viruses have a specific association with their animal vectors which may include leafhoppers, aphids, thrips, whiteflies, mealy bugs, mites and nematodes. It should be pointed out that most of these animals feed by piercing tissues with their mouthparts and not by biting them. Biting insects have limitations as vectors since biting not only damages leaves excessively but is a method most likely to succeed only with viruses that can be spread by mechanical inoculation. Once inside the cell, uncoating of the virus probably takes place in a similar fashion to animal viruses.

Arthropod-transmitted animal viruses

The infection of animals with *arboviruses* resembles in some ways the infection of plants by the feeding activities of virus-carrying insects. The word arbovirus is an abbreviation for 'arthropod-borne virus of vertebrates' and includes members of such diverse groups as the Rhabdoviridae, Bunyaviridae, Reoviridae (orbiviruses) and Togaviridae (alphaviruses and flaviviruses). These viruses are maintained in nature principally through biological transmission between susceptible vertebrate hosts by hematophagous arthropods. The virus cannot be transmitted immediately to a new host following feeding on a viraemic host since the virus must multiply in the vector before transmission can occur. However, while virologists first learned of these viruses as a result of people becoming infected it seems likely that man is incidental to the natural insect-vertebrate relationship. The natural vertebrate host of these viruses, which occur mostly in the tropics, is commonly a rodent, a monkey or a bird. Infection of man appears to be accidental and may well break the zoonosis, or chain of transmission.

PREVENTION OF THE EARLY STAGES OF INFECTION

One of the goals of studying the cell-virus relationship is to develop methods of aborting viral infections, particularly those of man and his domestic animals, and how better than to prevent virus from attaching and penetrating. Cellular receptors are likely to be proteins or glycoproteins which perform functions vital to the normal metabolism of the cell and serve only incidentally the needs of the virus. Thus these cannot be attacked without endangering the cell itself. As to the viruses, little is known of the chemistry of the coat proteins so the search for anti-viral drugs which inhibit adsorption and penetration can only be achieved by mass screening techniques. So far such screening procedures have only uncovered one drug suitable for clinical use. This is amantadine (Fig. 4.9) which prevents some step in the penetration of influenza but its exact mode of action remains unclear.

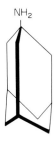

Figure 4.9 Structure of 1-amantadine.

For the protection of plants it should be apparent that the best preventive measure is a reduction in the number of appropriate vectors but this is not necessarily an easy task. Bacteriophages can also cause problems since they are capable of infecting certain organisms of industrial fermentations resulting in lysis and loss of product. Obviously the best measure to adopt here is the use of resistant strains and the simple measure of reducing the concentration of divalent cations since the latter are frequently important for phage adsorption.

5 The process of infection: IIA the Baltimore classification

Viruses exhibit an incredible diversity of morphologies, nucleic acid structure, mode of infection, regulation of development, etc. In such circumstances it might be thought impossible to uncover any unifying concept which would simplify a discussion of the process of replication. However, Baltimore has proposed an elegant classification of all viruses based on the mode of gene replication and expression. In this classification, mRNA is assigned a central role since protein synthesis takes place by the same mechanism in all cells. All viruses are then divided into groups, assignment to a group being determined by the pathway of mRNA synthesis (Fig. 5.1).

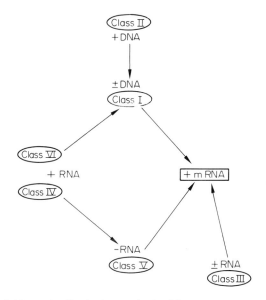

Figure 5.1   The Baltimore classification (see text for details).

In order to maintain unity, all mRNA is designated as ' + ' RNA. Strands of viral DNA and RNA which are complementary to the mRNA are designated as '-' and those which are non-complentary as ' + '. Using this terminology, six groups of viruses can be distinguished:

Class I   consists of all viruses which have a double-stranded DNA genome. In this class the designation of ' + ' and '-' is not meaningful since different mRNA species come from different strands.
Class II   consists of viruses which have a single-stranded DNA genome of the same sequence as mRNA. As yet no viruses with single-stranded DNA have

75

been found in which the genome is complementary to the mRNA; these would obviously go in a new class. Before the synthesis of mRNA can proceed, the DNA must be converted to a double-stranded form.

Class III   consists of viruses which have a double-stranded RNA genome. All known viruses of this type have fragmented genomes but mRNA is only synthesized on one strand of each fragment.

Class IV   consists of viruses with a single-stranded RNA genome of the same sequence as mRNA. Synthesis of a complementary strand precedes synthesis of mRNA.

Class V   consists of viruses which have a single-stranded RNA genome which is complementary in base sequence to the messenger RNA.

Class VI   consists of viruses which have a single-stranded RNA genome and which have a DNA intermediate during replication. At least some members have virion RNA and mRNA of the same polarity.

Because the replication of class II viruses involves a double-stranded DNA intermediate they will be considered along with class I viruses in Chapter 6. For similar reasons, the replication of classes III, IV and V will be discussed together in Chapter 7. Amongst class VI viruses are tumour viruses and their replication will be discussed in Chapter 15 when we consider the mechanism of tumour formation.

Finally, it must be emphasized here that classifications of viruses based on other criteria are equally valid and frequently far more useful. A weakness of Baltimore's scheme is that it takes no account of many important properties. Thus, classified together in class I are bacteriophage T2 and smallpox virus, agents which are totally dissimilar in structure and biology. The reader should refer to other texts (see Suggested Reading on page 240) and Chapter 18 to see how the problems of classification have been approached by 'non-molecular' virologists.

6 The process of infection: IIB the replication of viral DNA

The basic mechanism whereby a DNA molecule is duplicated *in vivo* is the same regardless of whether the DNA is of cellular or viral origin. Indeed, because of the ease with which viral DNA molecules can be manipulated *in vitro* they have often been the substrate preferred by biochemists attempting to unravel the mysteries of DNA replication. Despite the fact that it is two decades since the first DNA polymerase was isolated we still know little about the detailed mechanism. At first sight this is surprising because the replication of DNA would appear to be a simple process; a polymerase would traverse the molecule and make two daughter strands using the parental strands as templates (Fig. 6.1).

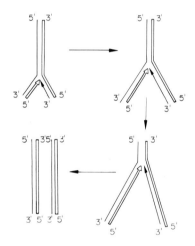

Figure 6.1 Simple model for DNA replication. In this and subsequent diagrams the arrowheads indicate the positions of the growing points. Note that in the model shown above, one strand is synthesized in the 3′ to 5′ direction and the other in the 5′ to 3′ direction.

This simple model for DNA replication implies that each of the daughter molecules contains one parental strand and one newly-synthesized strand, i.e. that DNA replication is semi-conservative. Proof of this was obtained from the now classical Meselson-Stahl experiment and represents one of the few indisputable facts concerning DNA replication!

Problems with polymerases

Since the two strands of a DNA duplex are anti-parallel, a second implication of our model for DNA replication is that one of the daughter strands is synthesized in the 3′ to 5′ direction and the other in the 5′ to 3′ direction (Fig. 6.1). However, of the many cellular and viral-specified DNA polymerases which have been purified none has the capacity to synthesize DNA in the 3' to 5' direction. All are incapable of adding monodeoxyribonucleotides to 5' hydroxyl termini. One solution to the problem is for synthesis to proceed in the 5' to 3' direction along one parental strand, the *leading* strand, and for discontinuous 5' to 3' synthesis to occur along the other, or *lagging*, strand (Fig. 6.2). The fragments produced by discontinuous synthesis could ultimately be tied together by a ligase.

Evidence supporting discontinuous synthesis has been obtained by Okazaki. A culture of *E. coli* which had been infected with phage T4 was pulsed with radioactive thymidine and the resulting labelled DNA examined by velocity sedimentation in a sucrose gradient. Immediately after the pulse most of the label was found in DNA fragments (called 'Okazaki fragments') approximately 1000-2000 nucleotides long. By one minute later the radioactivity had been chased into material which sedimented much faster, as would be expected if the fragments had been covalently linked together by ligase action.

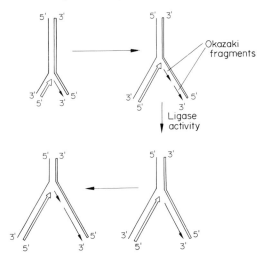

Figure 6.2 Model for discontinuous synthesis of DNA. Both strands are synthesized in a 5' to 3' direction but only one strand is synthesized continuously. Experimental evidence (see text) suggests that both strands may be synthesized discontinuously.

A second problem with our simple model is the apparently perverse inability of known DNA polymerases to start DNA chains *de novo*. Such new chain starts are required at frequent intervals on the lagging side of the fork. A solution to our problem is to invoke the action of RNA polymerase which does not require a primer since its basic function is to start polynucleotide synthesis. This enzyme could synthesize a short RNA primer on the lagging strand and this could be extended by DNA polymerase and ligated to the previously synthesized Okazaki fragment. In support of this model is the observation that the antibiotic rifampicin, which inhibits the enzyme RNA polymerase, is known to inhibit the conversion of single-stranded M-13 DNA to the double-stranded form. No such inhibition is observed in an *E. coli* mutant with a rifampicin-resistant RNA polymerase. Furthermore, many DNA polymerases do not discriminate against extending a chain with a ribonucleotide at the growing end. Clearly, the synthesis of each Okazaki fragment will require a short RNA primer and this primer has to be excised after synthesis of the fragment and the space left filled by extension of the next fragment. Experimentally it is found that remnants of the RNA primer disappear so rapidly that one suspects that primer degradation may be tightly coupled to primer utilization. Both processes might be catalysed by the same enzyme complex, thereby maximizing the efficiency of primer removal.

Why the requirement for an RNA primer?

The model of DNA replication shown in Fig. 6.3 is so much more complex than that shown in Fig. 6.1 that one wonders if it is possible that nature has been foolish and wasteful in engineering DNA replication systems. This is unlikely and a rationale for the complexity is provided by the observation that the fidelity of DNA replication is such that only one mistake is made in 10^9-10^{10} base-pair replications. Error-free replication arises from the ability of DNA polymerases to 'proof-read' the DNA which they have just synthesized. For example, T4 and *E. coli* DNA polymerases have a very strong requirement for a Watson-Crick

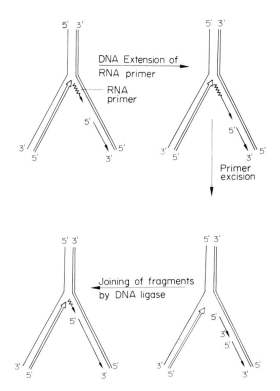

Figure 6.3 Involvement of an RNA primer in synthesis of Okazaki fragments.

base-paired residue at the 3'-OH primer terminus to which nucleotides are being added. When confronted with a template-primer with a terminal mismatch, these polymerases make use of their built-in 3' → 5' exonuclease activities to clip off unpaired primer residues by hydrolysis until a base-paired terminus is created. As a result, these polymerases will efficiently remove their own polymerization errors. This self-correcting feature allows DNA polymerase to select for the proper template base-pairing of each added nucleoside triphosphate in a separate backward reaction, in addition to its strong selection for base-pairing of nucleoside triphosphates during the initial polymerization. If DNA polymerase had to initiate new DNA chains in the absence of a primer this proof-reading function would not be operative at the initiation regions. By contrast, RNA polymerases need not be self-correcting, in as much as relatively high error rates can be tolerated during transcription. Indeed, it seems reasonable to suggest that the use of ribonucleotides in the synthesis of the primer was an evolutionary step forward for it automatically marks these sequences as 'bad copy' to be removed.

The above line of reasoning also suggests an explanation for the failure to find a DNA polymerase which adds deoxyribonucleoside 5'-triphosphates in such a way as to cause chains to grow in the 3' → 5' direction, as depicted in our simple model of DNA replication shown in Fig. 6.1. With such a 3' → 5' polymerase the growing 5' chain end, and not the incoming mononucleotide, would carry the triphosphate activation. Thus mistakes in polymerization could not be hydrolysed away without a special enzymatic system for reactivating the base 5' chain end thus created.

The Mechanism of Priming

The small DNA phages M13, G4 and φX174 have proved to be excellent models for studying the priming of DNA synthesis, for their single-stranded (SS) circular genomes have no free 3' OH termini. When the conversion of M13 SS DNA to the double-stranded replicative form (RF) was followed in crude extracts, two stages could be recognized. The initial stage involved RNA synthesis and produced primed SS molecules which could be isolated. Such primed SS molecules were converted to RF molecules in a second stage in the absence of ribonucleotide triphosphates even if rifampicin were present. Resolution and purification of the components needed for the conversion of M13 SS to RF disclosed requirements for five proteins (Fig. 6.4). RNA polymerase binds to M13 SS DNA, but only in the presence of a DNA unwinding protein, and synthesizes a short RNA primer. The RNA primed SS DNA is extended to form RF II molecules by the *E. coli* DNA polymerase III holoenzyme. DNA polymerase I then excises the RNA primer and fills the gap thus created and DNA ligase joins the two ends of the newly synthesized polynucleotide.

The enzyme system active in initiation of G4 SS to RF conversion is distinct from that for M13 since it is unaffected by rifampicin and thus does not involve RNA polymerase. Rather, it has an absolute requirement for the product of the *E. coli dna*G gene which is a protein of mol. wt. 61 000. In the presence of G4 SS DNA and the DNA unwinding protein, the *dna*G protein catalyses the incorporation of the 4 ribonucleotide triphosphates into short RNA primers. Such RNA

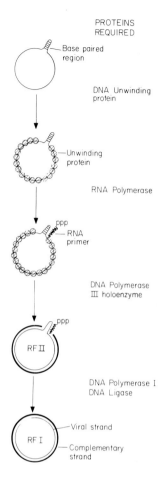

PROTEINS
REQUIRED

Base paired region

DNA Unwinding protein

Unwinding protein

RNA Polymerase

ppp
RNA primer

DNA Polymerase III holoenzyme

ppp

RF II

DNA Polymerase I
DNA Ligase

RF I

Viral strand

Complementary strand

Figure 6.4 Proteins required for the conversion of M13 single-stranded DNA to replicative form I.

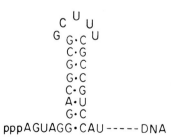

Figure 6.5 Sequence of the RNA primer required for the conversion of G4 single strands to replicative forms.

81

primers are transcripts of a unique segment of the G4 template and they have been isolated and sequenced (Fig. 6.5). The sequence suggests the presence of a G-C rich duplex region of 8 base-pairs which is believed to be present in the G4 chromosome *in vivo*. Such a duplex region may not be coated with DNA unwinding protein and might serve as a recognition signal for *dna*G protein action.

Despite the fact that ϕX174 and G4 are morphologically identical and possess DNA molecules of similar size, the two do not prime DNA synthesis in exactly the same way. Rather, the ϕX174 initiation system is much more complex. The products of the *E. coli dna*B and *dna*C genes, as well as proteins i and n (whose genetic loci are unknown), interact with ϕX174 SS DNA complexed with the unwinding protein. Only when this replication intermediate has been formed will the *dna*G protein synthesize the primer. The fact that phages M13, G4 and ϕX174, all with circular single-stranded genomes of mol. wt. 1.6×10^6, prime DNA synthesis in different ways indicates different evolutionary histories.

Studies on the replication of polyoma virus DNA have provided some of the most convincing evidence that nascent DNA fragments are primed by RNA *in vivo*. The replication of polyoma virus DNA, which is a circular duplex of mol. wt. 3×10^6 daltons and similar in size to ϕX174 RF, proceeds discontinuously with synthesis of small DNA fragments, 100-140 nucleotides long. Such fragments synthesized in mouse cells, or in nuclei isolated from them, appear to have RNA at their 5' ends, but with no sequence specificity at the covalent linkage of RNA to DNA. The RNA primer appears to be a decanucleotide with ATP or GTP at the 5' terminus.

Genetic and biochemical characterization of bacteriophage T4 replication

In 1963 Edgar and Epstein presented their now classical analysis of the conditional lethal mutants of T4 (see p. 13). These mutants, which mapped into many different complementation groups or genes, helped identify six protein components of the DNA replication system—the products of genes 32, 41, 43, 44, 45 and 62. These mutations are characterized by the fact that infection under non-permissive conditions gives rise to little or no DNA synthesis, even though the deoxyribonucleoside triphosphate precursors are available. The first of the T4 replication gene products to be identified and extensively characterized was the product of gene 43, T4 DNA polymerase. Attempts to decipher the function of the other gene products have relied upon the generation of new biochemical methods such as DNA cellulose chromatography. This method assumes that many of the proteins which function in association with intracellular DNA will recognize purified DNA as a substrate and bind tightly to it *in vitro*. Indeed, when infected cell extracts were chromatographed on DNA columns, about 20 different T4-specified proteins were subsequently resolved by polyacrylamide gel electrophoresis. By comparing these DNA binding proteins from mutant and wild-type infections, it was possible to use genetics to identify protein species as the product of particular bacteriophage genes. In this way, for example, the gene 32 protein was identified, isolated, and subsequently characterized as a DNA unwinding protein.

The gene 44 and 62 proteins are normally isolated as a tight complex which appears to contain four molecules of the gene 44 protein and two molecules of the gene 62 protein. The other four replication proteins contain only a single type of sub-unit, although the gene 32 protein appears to exist as an aggregate, and the gene 45 protein as a dimer. The gene 41 protein appears to be a monomer as judged by its sedimentation rate but there are indications that in its functional stage it may be a dimer. The functions of the genes 44, 45 and 62 products are not known except that they act co-operatively *in vitro* to stimulate T4 DNA polymerase. The product of gene 41 may function in the synthesis of the RNA primer required for initiation of DNA synthesis. When all six proteins are added to T4 DNA *in vitro,* DNA synthesis is rapidly initiated but only if ribonucleoside triphosphates are present as well as deoxyribonucleoside triphosphates. It is therefore not clear if synthesis on the leading strand is continuous or discontinuous.

Problems with primers

The requirement for a primer causes a further complication which we have so far ignored. If synthesis in the 5′ to 3′ direction is initiated with an RNA primer which is later digested away, there exists no mechanism for filling the gap left by the primer at the 5′ end of the leading strand. To fill this gap would require 3′ to 5′ synthesis and we know that this cannot occur. When the replicating fork reaches the other end a similar problem arises with the lagging strand (Fig. 6.6).

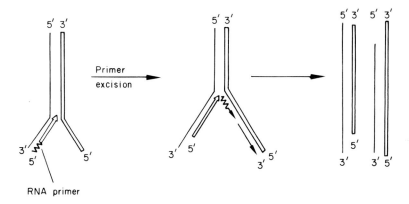

Figure 6.6 Formation of two daughter molecules with complementary single-stranded 3′ tails.

The net result is the synthesis of two daughter molecules each with a 3′ single-stranded tail. If such molecules were to undergo further rounds of replication the result would be smaller and smaller 3′ tailed duplexes. One solution to this problem is concatemer formation.

Concatemer formation

It will be recalled from Chapter 3 that many viruses have terminally repetitious genomes and that exonuclease treatment of such genomes results in the production of 'sticky ends' enabling concatemers to be produced. Consequently, if a pair of tailed molecules are produced during replication, they too should be able to form concatemers. Concatemer formation would be independent of the length of the terminal repetition and the replication tail. If the tails were longer than the redundant region then gaps would be produced which could only be filled with the aid of a DNA polymerase and a DNA ligase. If the tails were shorter than the redundant stretch then concatemer formation would be the result of exonuclease and ligase action (Fig. 6.7).

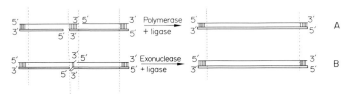

Figure 6.7 Dimer formation by molecules with complementary tails. (A) Molecule in which the tails are longer than the redundant region. (B) Molecules in which the tails are shorter than the redundant region.

Formation of unit length molecules would occur by the reverse process to concatemer formation. Two specific enzymatic breaks would be introduced, one in each strand but at opposite ends of the repetitious sequence (Fig. 6.7). Creation of such cuts would probably not cause separation of the two halves since there would be enough hydrogen bonds to hold them together. If the cuts were introduced such that the longer of the two chains has a 5′ terminal then a DNA polymerase could add deoxyribonucleotides to the 3′ hydroxyl end. This would displace a 5′ ended tail. When the growing 3′ ends meet somewhere in the middle of the redundant section there would no longer be any hydrogen bonds holding the two halves together; they would thus separate. The two halves could either reform into a hydrogen-bonded concatemer or else be converted to a mature terminally repetitious molecule by the continuing action of DNA polymerase (Fig. 6.8).

Experimental evidence in favour of the above model comes from a study of T7 replication. If T7 infected cells are pulsed with ^{3}H-thymidine, all of the label is found in phage DNA since host DNA synthesis ceases shortly after infection. When analysed in sucrose gradients most of this label sediments faster than DNA from purified T7. Measurement of the sedimentation rate of labelled molecules indicates that they are dimers, trimers, tetramers etc., of T7 DNA. Furthermore, the label can be chased from such concatemers into molecules the same size as mature phage DNA.

The smaller concatemers have been subjected to intensive analysis and the results support the model just outlined. Denaturation of dimers and trimers and sedimentation of the resultant chains in alkali revealed unit length molecules as expected. Chromatographic separation of sheared monomers, dimers and trimers showed that they have single-stranded regions spaced at intervals corresponding to the length of a mature phage DNA molecule. Furthermore, annealing experiments with exonuclease-treated phage DNA demonstrated that single-stranded regions correspond to the ends of a mature molecule. Finally,

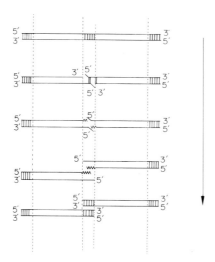

Figure 6.8 Conversion of dimer to two unit length molecules with terminally redundant ends. Terminally redundant regions of the molecule are represented by hatching.

studies with an endonuclease and an exonuclease specific for single-stranded DNA showed that both tails and gaps are present in the concatemers.

The model represented above should be applicable to any virus with a terminal repetition, e.g. the T-even phages, P22 and SP50. However, some of these viruses have genomes which are circularly permuted and during the formation of 'phage lengths' of DNA the cutting nuclease apparently does not recognize any specific sequence. Instead, some spatial factor probably fixes the site of nuclease action. In the case of T4, empty head shells are assembled and then subsequently filled with DNA. Presumably when a headful has been packaged, a nuclease snips off the remaining unpackaged concatemer. This model is supported by the observation that abberant T4 particles with giant heads contain continuous lengths of DNA much longer than the usual genome.

THE REPLICATION OF CIRCULAR DNA

Both single-stranded and double-stranded circular DNA genomes have been found in viruses. Since the replication of single-stranded DNA always involves a double-stranded intermediate we shall first turn our attention to the replication of double-stranded circular genomes as exemplified by bacteriophage λ.

About 30 minutes after infection with λ, pulse-labelled DNA from infected cells can be separated into three classes by sedimentation—linear monomers, covalently closed circles, and concatemers. Most of the label is found in the concatemers and can be chased into linear monomers. What type of structure is the concatemer and how might it be generated? The two simplest concatemeric structures are linear multimers and multimeric circles and ways of generating such structures are shown in Fig. 6.9. Concatemers also may be generated by replication of a rolling circle (Fig. 6.9) in which the circle corresponds to a unit genome and the tail is a linear multimer. These concatemers are probably

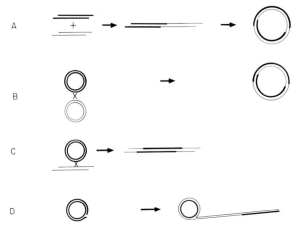

Figure 6.9 Models for the generation of concatemers. (A) End to end joining of monomers. Circular concatemers can be generated by joining the two free ends of the linear form. (B) Reciprocal recombination between circular monomers generates a circular dimer. (C) Reciprocal recombination between a circular monomer and a linear monomer generates a linear dimer. (D) Replication of a rolling circle generates a structure in which the circular duplex is monomeric in length but the tail may be of any length depending upon how long replication has proceeded.

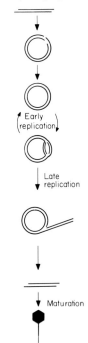

Figure 6.10 A model for the replication of bacteriophage λ DNA. Upon entering the cell the linear duplex is converted to an open circle and then to a closed circle. In early replication the circular template assumes a theta shape and replication proceeds bi-directionally. This generates circles which do not act as precursors for mature linear duplexes. During late replication there is a change in the mode of replication to a rolling circle which generates linear tails which serve as precursors for mature duplexes.

produced by replication for they are still formed when a recombination-deficient mutant of λ infects a recombination-deficient host cell, a procedure which ensures the absence of recombination functions.

Fig. 6.10 summarizes the replication cycle of λ. The DNA in the virion is a linear monomer which is converted to a closed circle after infection. During the period of early replication λ DNA replicates in a circular form, usually bi-directionally, to generate circular progeny. Late replication is initiated by the conversion of circular DNA to the rolling circle but the events responsible for this transition are not known. Replication of the rolling circle generates a concatemeric tail which is cleaved during packaging into molecules of the correct length and possessing cohesive ends. Lambda thus uses two strategies to overcome the problem of single-stranded tails at the 5′ end of the leading DNA strand. By replicating via a circular molecule early in infection, the gap formed by removal of the RNA primer is filled by the polymerase after circumnavigating the circle. Late in infection the synthesis of concatemers again prevents any problems arising by the removal of the terminal RNA primers.

Replication of single-stranded circular genomes

The genomes of a variety of animal and bacterial viruses consist of single-stranded circular DNA (see p. 44) but the only well-studied example is that of bacteriophage φX174. In all cases the infecting single-strand is converted to a double-stranded replicative form (RF) by the mechanism outlined on page 80. This parental RF then replicates to generate progeny RF but at present it is not clear whether this occurs via a theta structure or a rolling-circle since evidence in favour of both models has been presented. What is certain is that the leading strand is synthesized on the complementary strand of the RF while the Okazaki fragments are synthesized on the viral strand of the RF (Fig. 6.11). This has been shown in a number of different ways, the simplest being the observation that denaturation of RF synthesized in cells in the absence of DNA ligase releases short fragments from the complementary strand while the viral strands are of unit length.

The first suggestion that production of φX174 single strands might be accomplished by a rolling circle was made by Gilbert and Dressler in 1968 in their original and classic proposal of the model. It has since been shown that a pulse of ^{3}H-thymine given during the period of single-strand synthesis enters a fraction sedimenting heterogeneously in the range 16-30 S, compared with 24S for SS φX174 DNA. When intermediates of different sizes were denatured by alkali and the strands separated by buoyant density, the faster sedimenting intermediates proved to have the longer viral strands up to two unit genomes in length. Only the viral strand could be detected by the incorporation of a radioactive label during the period of single-strand synthesis, implying that the synthesis of complementary strands had ceased. As would be expected from denaturation of a rolling circle intermediate, complementary strands can be recovered, but in the form of unit length circles, and these can only be detected by an infectivity assay. Examination by electron microscopy of the replication intermediates

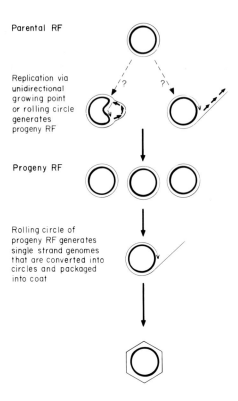

Parental RF

Replication via
unidirectional
growing point
or rolling circle
generates
progeny RF

Progeny RF

Rolling circle of
progeny RF generates
single strand genomes
that are converted into
circles and packaged
into coat

Figure 6.11 The replication of bacteriophage φX174. The encapsulated DNA strand is represented by the thin line.

found during the period of single-strand synthesis revealed the presence of the expected rolling circles molecules.

One of the features of the rolling circle model which remains to be elucidated is the reaction responsible for generating a circle from the linear single-stranded tail. There have been various suggestions that linear strands of φX174 DNA have an intrinsic ability to form circles, perhaps because of the presence of a palindromic sequence able to support formation of a hairpin. Unequivocal proof of this is lacking. Another feature which has to be explained is the mechanism whereby synthesis of the complementary strand is prevented. There is good evidence to show that formation of single strands is tightly coupled to synthesis of coat proteins. Perhaps the single-stranded tail is complexed with coat proteins as soon as it is synthesized thus preventing synthesis of the Okazaki fragments. Thus assembly of the phage particle and packaging of the DNA could take place simultaneously. However, this still does not explain how the ends of the linear molecule are sealed into a circle and it should be borne in mind that if cellular DNA ligase carries out this function there will be a requirement for a *double-stranded* region of the genome!

THE REPLICATION OF LINEAR SINGLE-STRANDED DNA

A linear single-stranded genome is found in both the autonomous and defective parvoviruses. The autonomous parvoviruses use host cell enzymes for replication and transcription and package a unique viral DNA strand. Defective parvoviruses such as the adeno-associated viruses (AAV) are entirely dependent on adenovirus co-infection for their own replication and package both plus and minus DNA strands in separate virions. The parvovirus genome contains terminal hairpins (Fig. 6.12) and this is an important feature for it provides a 3'-OH terminus at which DNA elongation can be initiated. However, the overall genome structure is different for the defective and autonomous parvoviruses (Fig. 6.12) and this provides an explanation for the observation that viruses of one group package both strands of DNA and the others do not.

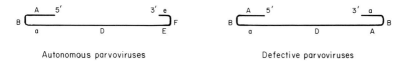

Autonomous parvoviruses   Defective parvoviruses

Figure 6.12 Schematic representation of the genome structure of autonomous and defective parvoviruses. In this and the subsequent two figures the letters A and a etc., represent complementary sequences.

Cells infected by either type of parvovirus accumulate double-stranded forms of DNA that appear to be intermediates in replication. A large fraction of such double-stranded molecules cannot be irreversibly denatured, suggesting that the two strands are covalently linked. Linear dimeric double-stranded DNA molecules that spontaneously renature have also been observed. The monomeric and dimeric double-stranded DNAs appear to involve plus and minus strands linked end to end: one of each in the monomer and four alternating plus and minus strands in the dimer. These observations suggested a model for parvovirus DNA replication.

Let us first consider the model for replication of an autonomous parvovirus (Fig. 6.13). The first step involves gap-fill synthesis to generate complement d of sequence D. This is followed by displacement synthesis and gap-fill synthesis, to copy the 5' terminal hairpin sequence aBA. The structure now undergoes rearrangement to form a 'rabbit-eared' structure which recreates the hairpin originally present at the 5' end of the parental genome. More important, it simultaneously creates a copy of this hairpin at the 3' end of the complementary strand and which can serve as a primer. The dimer length duplex is then completed by displacement synthesis. The resulting molecule comprises a single poly-nucleotide chain from which two viral genomes could be generated by endonuclease action. However, in the absence of endonuclease action, continued replication could occur to generate larger multimers (as shown in Fig. 6.13).

The dimers and multimers created by the process just described could serve as replicative intermediates from which progeny viral DNA would be excised by a similar method of displacement synthesis. A single-stranded break would be introduced at the 5' end of a genome within the concatemer. The 3'-OH terminus then acts as a primer for displacement synthesis of progeny

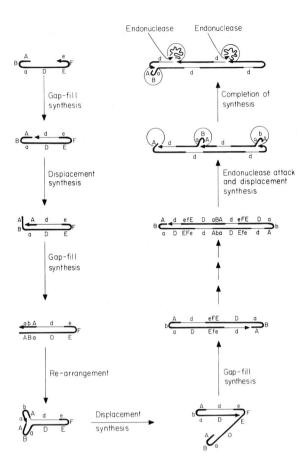

Figure 6.13 A scheme for the replication of an autonomous parvovirus (see text for details).

DNA strands, possibly driven by the packaging process. After the 3′ terminal sequence e had been displaced excision of the progeny genome could be completed by another site-specific endonuclease, resulting in the release of an intact virus particle and termination of the displacement synthesis. This model neatly accounts for two experimental observations. Firstly, duplex molecules with single-stranded tails have been observed in cells infected with minute virus of mice (MVM), an autonomous parvovirus. Secondly, no free single-stranded DNA has been detected in such infected cells.

An interesting feature of this model of replication concerns the single-stranded loop sequences B and F. Cleavage of a tetrameric replicative intermediate (see Fig. 6.13) would generate four different viral DNA molecules, i.e. molecules containing sequences B + F, b + F, B + f and b + f. The sequences B and F are known to be less than 100 nucleotides long and since only 60% of the genome is transcribed B and F probably represent non-coding sequences. This variability thus would not be any disadvantage to the virus. However, if the single-stranded loop sequences B and F had the potential to form a double-stranded palindrome, e.g.

$$- GGTACC - \rightarrow - GGTACC -$$
$$- CCATGG -$$

there would be no difference between sequences B and b or between sequences F and f. Thus all genomes would be identical.

A variation of the model just presented can be used to describe the replication of defective parvoviruses such as AAV (Fig. 6.14). The self-complementary terminal segments of AAV DNA single strands would fold back on themselves to form short duplexes. The 3′ end of one or both strands could then serve as a primer for synthesis of the complementary strand. The duplex so constructed would then be nicked near the hairpin and the 5′ ended segment, containing a full complement of terminal sequences, would then serve as a template for completion of the other strand by 5′ to 3′ synthesis. Because both ends of the duplex are identical, a consequence of the inverted repetition, both plus and minus strands could be generated by displacement synthesis after self-priming.

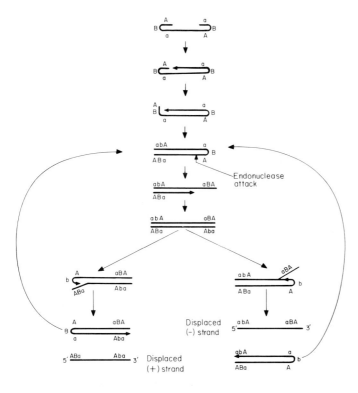

Figure 6.14 A scheme for the replication of a defective parvovirus (see text for details).

SYNTHESIS OF DNA CONTAINING MODIFIED BASES

You will recall from Chapter 3 that the DNA from certain bacteriophages contains modified bases. For example, 5-hydroxymethyl cytosine (HMC) replaces cytosine in T-even phage DNA, and some of these HMC residues are glucosylated. How is cytosine excluded from the DNA of these phages and how are the modified bases synthesized? The asking of these and other questions about the biosynthesis of the DNA of the T-even phages stimulated several laboratories to study the biochemistry of virus infection. The outcome was the identification of a whole series of new enzymes. Since these enzymes must be synthesized prior to the initiation of DNA replication, they make their appearance early in the infection process. In fact they constitute the bulk of the early proteins identified by Hershey (page 9).

Approximately four minutes after infection an activity appears in *E. coli* infected with T2 that cleaves dCTP and dCDP to dCMP. This appears to be the major mechanism by which dCTP is prevented from being incorporated into phage DNA. The major source of dCTP synthesis in uninfected and infected *E. coli* is via the reduction of CDP to dCDP and the conversion of dCDP to dCTP. Thus it seems reasonable that the phage-induced dCDP-dCTPase should attack both the di- and tri-phosphate. The function of this dCDP-dCTPase appears to be the prevention of accumulation of dCTP as a DNA polymerase substrate and also to maintain a reasonable pool of dCMP. The utilization of dCMP is via two reactions: deamination to dUMP which is a precursor of dTMP, and hydroxymethylation to dHMCMP. Hydroxymethylation of dCMP is a reaction unique to cells infected with T-even phages and requires tetra-hydrofolic acid. Hydroxymethylase activity is neither found in uninfected *E. coli* nor in cells infected with T5, whose DNA contains cytosine.

If an extract of a T-even infected *E. coli* is assayed for DNA polymerase, very little activity is found. However, if dCTPase is inhibited by fluoride, or dHMCTP is used instead of dCTP, DNA polymerase activity can be detected. When this

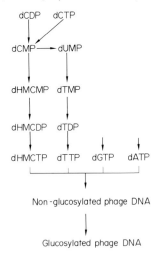

Figure 6.15 Biosynthesis of phage T2 DNA.

92

polymerase activity is purified it fails to be inhibited by antiserum prepared against *E. coli* DNA polymerase. Since this enzyme also has different template requirements it appears to be a new enzyme. Indeed conditional-lethal mutants of T4 have been isolated which produced a DNA polymerase inactive at restrictive temperatures. Failure of the T2-induced polymerase to utilize dCTP further ensures that cytosine is absent from phage DNA. However, this polymerase does not produce T2 DNA since the HMC residues are not glucosylated. Glucosylation is achieved through the appearance of yet another set of phage-specified enzymes which catalyse the transfer of glucose from UDP-glucose to the HMC residues in the DNA.

All of the reactions described above are outlined in Fig. 6.15. Similar enzymes are known to be responsible for the synthesis of the other modified bases found in phage DNA molecules.

HOST-CONTROLLED RESTRICTION AND MODIFICATION

The phenomenon of host-controlled restriction is most easily observed as the inactivation of bacteriophage following transfer from one bacterial strain to another. Thus phage λ grown on strain K12 (λ. K) may form plaques on strain K12 (P1) but with a much lower efficiency. Since λ. K eclipses equally well on both strains, multiplication of the phage must be restricted in strain K12 (P1). Those phages (λ. K, P1) that do survive restriction and are propagated on strain K12 (P1) are modified in such a way that they are no longer restricted by this host (Fig. 6.16). By labelling λ. K DNA with ^{32}P it has been shown that the

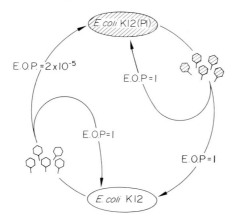

Figure 6.16 Restriction of phage λ by *E. coli* strain K12 (P1). The efficiency of plating is abbreviated to E.O.P.

reduced efficiency of plating on strain K12 (P1) is due to the degradation of phage DNA shortly after infection. Since no degradation occurs upon infection of strain K12, restriction must act at the level of DNA. As we shall see later, modification also acts at this level.

Most of the progeny issuing from strain K12 after infection with λ. K, P1 have lost the capacity to grow on K12 (P1). Apparently, those phages which can still grow on K12 (P1) have at least one conserved parental DNA strand. This was shown by growing λ. K, P1 phage in strain K12 in the presence of ^{32}P and then correlating the ability of the progeny to grow on K12 (P1) and their sensitivity to autoradiolysis by incorporated ^{32}P (Fig. 6.17). An aliquot of heavily ^{32}P-labelled

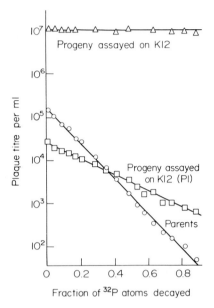

Figure 6.17 Correlation between ability to grow on K12 (P1) and sensitivity to autoradiolysis by ^{32}P incorporated in progeny phage particles produced in strain K12 after infection with ^{32}P-labelled λ. K, P1. See text for details.

λ. K, P1 was stored at 4°C and assayed from day to day. Plated on either K12 or K12 (P1) it gave the same plaque titres. Another aliquot was allowed to grow for one cycle in K12 bacteria in non-radioactive medium. Dilutions of this lysate were also stored at 4°C and assayed on K12 and K12 (P1). Since strain K12 permits the growth of all progeny the titres obtained were independent of the time of storage, an indication that a large majority of the particles contained little or no ^{32}P. Strain K12 (P1) permits only the growth of phage with the K, P1-specific modification and the assays showed that all such particles contained ^{32}P since their activity decreased with time of storage. The slope of the curve, roughly half that observed with the parental phage stock, indicates that the λ. K, P1 progeny phage contained half as much ^{32}P as its parents.

Modification is usually very efficient, affecting all phage particles in the yield, but the efficiency can be affected by the environment. If λ is grown in a methionine-requiring cell in a medium lacking methionine, many of the progeny are not modified. Modification can be restored by adding methionine to the medium. Since this result is not observed after starvation for any of a number of other amino acids it can be concluded that methionine plays a key role in modification. It is possible to modify DNA *in vitro* using *E. coli* extracts and S-adenosylmethionine as methyl donor. Such extracts transfer ^{3}H-labelled methyl groups from S-adenosylmethionine to unmodified DNA, but not to modified DNA. The product of the reaction has been identified as 6-methyl-aminopurine. Thus modification consists of enzymatic methylation of adenine.

Breakdown of unmodified DNA calls for the presence of specific nucleases in strains exhibiting restriction and recently a number of such enzymes have been purified. All degrade unmodified DNA but not modified DNA. The number of restriction sites for some molecules is known and statistical considerations suggest that such sites can be fully determined by 6 to 9 base-pairs. Furthermore, sequence analysis of the targets of several restriction enzymes shows that they have palindromic sequences (Fig. 6.18) with scissions being introduced at the same point on either strand. It should be pointed out that a consideration of the mode of action of this enzyme lends credence to the hypotheses, advanced earlier during our discussion of DNA replication, for the introduction of staggered nicks into DNA molecules.

Enzyme	Source	Sequence recognized
*Eco*R I	*Escherichia coli* strain RY13	5′ GAATTC 3′ CTTAAG
Hind III	*Haemophilus influenzae* strain Rd	5′ AAGCTT 3′ TTCGAA
Sal I	*Streptomyces albus* strain G	5′ GTCGAC 3′ CAGCTG
Taq I	*Thermus aquaticus* strain YTI	5′ TCGA 3′ AGCT
Hae III	*Haemophilus aegyptius*	5′ GGCC 3′ CCGG

Figure 6.18 Palindromic DNA sequences recognized by restriction enzymes. The arrows indicate the point at which scissions occur and the asterisks indicate the bases modified by methylation.

As yet, restriction and modification have not been described for animal virus systems. However, they do provide an attractive system for the study of protein-nucleic acid interactions. Furthermore, many restriction enzymes are now used extensively in sequencing DNA molecules and in the construction of novel DNA molecules.

DEPENDENCE AND INDEPENDENCE AMONG DNA VIRUSES

The autonomy of viruses as regards the replication of their DNA varies between wide limits and is a function of the size of the viral genome. At one end of the scale are viruses such as the T-even phages and the poxviruses with genomes of molecular weight 1-2 $\times$ 10^8 daltons, corresponding to 100-200 genes. Such viruses probably require little from their host cells other than an enclosed environment, protein synthesizing machinery and a supply of amino acids and deoxyribonucleotide triphosphates. Some may not even require this much for we have already seen how T2-specified proteins are involved in the biosynthesis of modified bases. Other viruses such as herpes simplex and vaccinia specify new thymidine kinase activities and there are reports of new transfer RNA species appearing in infected cells.

At the other end of the scale are viruses such as ϕX174 and minute virus of mice (MVM) whose genomes (mol. wt. 1.7 $\times$ 10^6 daltons) cannot specify more than 10 proteins. Since there may be as many as four different coat proteins, not many genes are left to code for functions essential to replication. Such viruses probably rely on the host not only for nucleic acid precursors but also for polymerases, ligases, nucleases, etc.

One approach which can be taken to ascertain the contributions of the host and the virus towards viral DNA replication is to isolate host cell mutants which no longer permit viral replication. This approach has largely been confined to the study of phage-bacteria relationships and usually involves investigating the ability of a phage to grow in a known bacterial mutant. For example, most coliphages grow in mutants lacking DNA polymerase I, but P2 does not. Obviously a functional DNA polymerase I is a prerequisite for P2 replication. Again, T4 can replicate in mutants lacking DNA polymerase III, but ϕX174 cannot.

Many instances are known where animal viruses fail to replicate in a particular cell line although they are capable of initiating infection. For example, human adenoviruses can infect monkey kidney cells but fail to grow in them. Curiously enough, co-infection of the cells with SV40 permits adenovirus replication through the formation of progeny adenovirus—SV40 hybrids. These have a variable amount of the adenovirus and SV40 genomes covalently linked.

In other cell lines the adenoviruses themselves can act as 'helpers'. Many adenovirus preparations contain a contaminating virus which is much smaller. This virus, adeno-associated virus, has a single-stranded genome of mol. wt. 1.7 $\times$ 10^6 daltons, hence the coding potential of this genome is identical to that of ϕX174 or MVM. We would thus expect AAV to replicate independently of the parent adenovirus but this has not been observed. Perhaps AAV can replicate independently in some other type of cell. Alternatively, the virus may lack some essential function which only adenovirus can provide.

A satellite virus of a different type has recently been described. This is phage P4 which is a satellite of phage P2. Unlike most satellite viruses this phage can replicate its nucleic acid in the absence of helpers. Instead, it lacks all known genes for phage morphogenesis and its DNA is packaged in a head composed of helper phage proteins.

7

The process of infection
IIC RNA synthesis by RNA viruses

The synthesis of RNA by RNA viruses involves (1) replication which is defined as the production of progeny virus genomes and (2) transcription which is the production of RNA complementary to the genome. Since RNA genomes may be either + (e.g. class IV) or − (class V) sense, transcription is not always synonymous with the synthesis of mRNA as it is in DNA viruses.

SYNTHESIS OF THE RNA OF CLASS IV VIRUSES

Just as there are several ways in which a single-stranded DNA molecule could be replicated, there are several ways in which a single-stranded RNA molecule might replicate. Since most RNA viruses can replicate in the presence of inhibitors of DNA synthesis it is unlikely that a DNA intermediate is involved. This is not true, however, for the Retroviridae and their replication will be discussed in Chapter 15.

When RNA is extracted from infected cells, an RNase resistant fraction can be isolated which is probably double-stranded RNA. This RNase resistant material also bands in Cs_2SO_4 at the position expected for double-stranded RNA and exhibits a steep thermal denaturation profile confirming its double-stranded nature. This double-stranded form, also called RF by analogy with the $\phi X174$ system, was first discovered with encephalomyocarditis (EMC) virus and has since been found in all the systems studied.

When RNA is extracted from infected cells and analysed in a sucrose gradient or, better still, by polyacrylamide gel electrophoresis, a third kind of RNA molecule is found (Fig. 7.1). Mild RNase treatment causes this new form to

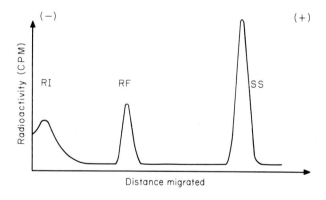

Figure 7.1 Separation of radioactive RNA components from virus-infected cells by polyacrylamide gel electrophoresis.

sediment as double-stranded RNA while it is completely degraded with more concentrated RNase. This suggests that it consists of a double-stranded molecule with single-stranded tails. Mild treatment of the replicative intermediate (RI), as this new form of RNA is called, with RNase removes the 'tails' leaving a double-stranded core (Fig. 7.2B,C). Concentrated RNase degrades the continuous template resulting in completely fragmented RNA.

As yet the role of RF and RI in the replication of viral RNA is not clear since these two classes of molecules appear to have different roles in the various systems studied. We probably know most about the replication of bacteriophage Qβ and this is the system which we shall discuss most fully. However, the evidence from other systems will be introduced where this differs significantly from that obtained with Qβ.

Both the RF duplex and the extensively hydrogen bonded RI of Qβ and other RNA phages are formed only after deproteinization (Fig. 7.2A,B). RF probably

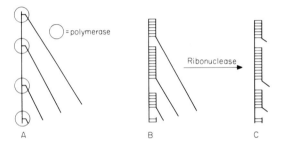

Figure 7.2   Proposed structure for a molecule of replicative intermediate before deproteinization (A), after deproteinization (B) and the effect of treating deproteinized RI with RNase (C).

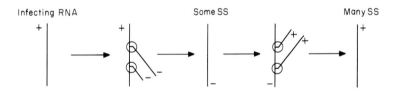

Figure 7.3   Synthesis of QβRNA. ○ = polymerase.

results from the annealing of free + and − strands during deproteinization. Thus during RNA synthesis the region of hydrogen bonding is confined to the sequences covered by the replicase and the enzyme makes and breaks hydrogen bonds as it proceeds. In other words, the RF and RI structure as we know them after RNA extraction are artefacts and are not concerned in RNA synthesis which proceeds via a template to which nascent RNA strands are held by the polymerase (Fig. 7.3).

In vivo the infecting QβRNA first directs the synthesis of the viral polymerase which together with host components uses the infecting RNA as a template for the synthesis of − strands. Then the − strand is copied into + strands. Production of + strands exceeds − strands by tenfold, possibly because + strands are removed as a source of template by being packaged into virus particles whereas − strands remain available as template continuously.

The synthesis of phage RNA *in vitro*

The synthesis of $Q\beta$ RNA has proven to be a useful reaction for studying the replication of RNA since the product of the *in vitro* reaction is infectious, the most stringent test possible for the fidelity of this system. Early studies on RNA replication *in vitro* were carried out with extracts of infected cells but purification of an RNA dependent RNA polymerase (also called RNA synthetase or RNA replicase) has now been achieved. The complete polymerase has a molecular weight of around 287 000 daltons and consists of one virus-coded and four host-coded polypeptides (Table 7.1).

Table 7.1 $Q\beta$ polymerase function and structure

Function	Source	Mol. wt. $\times 10^{-3}$
Binding to + strand	ribosomal protein S1	70
Initiation	elongation factor Tu	45
Initiation	elongation factor Ts	35
Chain elongation	$Q\beta$ encoded	65
− strand synthesis	ribosome-associated (hexameric) protein	72

The $Q\beta$ polymerase shows great specificity for templates; only $Q\beta$ RNA, the strand complementary to $Q\beta$ RNA and 6S RNA present in $Q\beta$ infected *E. coli* are known to undergo replication in the reaction catalysed by this enzyme. Recognition of the template must involve more than just the primary structure of a limited region of the RNA since fragments of RNA do not have template activity.

In vitro synthesis of the − strand was first detected by the appearance of an RNase-resistant product after deproteinization. As would be expected, the complementary strand is synthesized early in the reaction before progeny RNA is detected, and is correlated with the loss of infectivity of the template virus RNA molecules and the appearance of non-infectious RNase-resistant RNA in deproteinized reaction mixtures.

Although the RNA-dependent RNA polymerase is capable of synthesizing the − strand from viral RNA, the predominant product of the *in vitro* reaction is + strand RNA as demonstrated by the infectivity of the product and by hybridization. Thus the reaction is markedly asymmetric with respect to the synthesis of complementary and viral strands. The most obvious explanation that there are two enzymes involved—one to synthesize the complementary strand and the other to use this strand as a template for synthesis of new viral strands—is not borne out in fact as there is only one phage gene coding for an RNA polymerase, and furthermore, the phage genome is not sufficiently large to code for a second polymerase as well as the other phage-specified proteins.

Recent experiments show that the synthesis of − strands requires the viral polymerase in conjunction with the host proteins shown in Table 7.1 whereas a tetramer of the viral polymerase directs the synthesis of + strand RNA.

The Q*β* *in vitro* system has contributed to our understanding of the roles of RF and RI *in vivo* as explained above; for example, purified polymerase from phage-infected cells cannot use RF as template. Structures resembling the double-stranded RF and partially double-stranded RI found *in vivo* after deproteinization are also products of the *in vitro* reaction after deproteinization. Furthermore, in pulse-chase experiments, radioactivity in the RI could be chased into viral RNA. Thus, it has been suggested that the RI is a direct precursor of viral RNA *in vitro*. However, no one has yet been able to construct an active complex of purified enzyme and these double-stranded structures. The enzyme neither initiates synthesis with these molecules nor utilizes the nascent strand as primer for the continuation of synthesis. These observations support the view that the replicative template is not a double-stranded structure but rather a free complementary strand base-paired with nascent Q*β* RNA solely in the region of the polymerase.

The evolution of an RNA molecule in vitro

Regardless of the failure of studies on *in vitro* replication to provide an answer to the mechanism of *in vivo* replication, they have led to a very interesting series of experiments. Since the RNA instructs the *in vitro* replicative process, an opportunity is provided for studying the evolution of a self-replicating molecule outside a living cell. In the test-tube the RNA molecules are liberated from many of the restrictions of the complete virus life-cycle and the situation mimics a pre-cellular evolutionary stage when environmental selection presumably operated directly on the genetic material rather than on the gene product. The only restraint imposed is that they retain whatever sequences are necessary for the polymerase to recognize the RNA molecule. With this in mind, Spiegelman and his collaborators designed an experiment to determine what will happen to the RNA molecules when the only demand made on them is that they replicate as fast as possible. Since the product of the reaction of polymerase with RNA in itself acts as a template and because longer RNA chains take longer to be replicated, molecules would gain an advantage by discarding any unneeded genetic information to achieve a smaller size and therefore a more rapid completion. For example, suppose the polymerase were presented with an equal mixture of whole molecules and half molecules of Q*β*RNA. Because it takes half the time to replicate the half molecules compared to the whole molecules, there would soon be an excess of half molecules over whole molecules. Since there are now more half molecules than whole molecules, more half molecules would be used as templates leading to an even greater excess of half molecules. In this way there is rapid selection for the shorter molecules. Spiegelman and his collaborators prepared a reaction mixture containing the polymerase Q*β*RNA to act as template, and the appropriate nucleoside triphosphates. After incubation the mixture was diluted 12½ times into another tube containing the same amount of polymerase and triphosphates but *no* Q*β*RNA. This incubation and dilution sequence was repeated 75 times but whereas the dilution factor was kept constant, the incubation time was reduced gradually. After the 75th transfer, an RNA molecule was isolated which retained only 17% of the original RNA sequence. Furthermore, during these experiments a 'palaeontological' record was obtained by freezing each reaction tube until there was time to examine its contents. In this way it was observed that the size of the RNA molecule gradually decreased during the selection experiments.

The possibility now arises that qualitatively distinguishable phenotypes could be exhibited by a nucleic acid molecule under conditions in which its information is replicated but never translated. That this is indeed the case has been shown by Spiegelman and his co-workers who have been able to select variant molecules which replicate faster than input molecules in the presence of lower than usual concentrations of nucleoside triphosphates or in the presence of inhibitors such as tubercidin triphosphate (TuTP). TuTP is an analogue of ATP which inhibits the synthesis of Qβ RNA *in vitro*. It cannot completely replace ATP but is incorporated into growing RNA chains. A variant of QβRNA has been isolated which has an *in vitro* doubling time of 2 minutes in the presence of TuTP whereas the RNA molecule from which it was derived only doubles every 4.1 minutes; in the absence of TuTP both molecules have a doubling time of 1 minute.

These variant molecules which have been isolated from QβRNA combine a high affinity for replicase with a rapid growth rate and may also possess resistance to a chemotherapeutic agent (TuTP). The isolation of similar molecules from the RNA animal viruses could provide us with new chemotherapeutic agents.

The synthesis of poliovirus RNA

The mechanism by which poliovirus RNA is synthesized appears to be similar to that of Qβ. However, there are certain notable differences. First, the RF of poliovirus is more stable than that of the RNA coliphages and this may explain the existence of recombination only in the former. Secondly, RF molecules isolated from cells infected with poliovirus are infectious whereas those from Qβ infected cells are not. There is currently no explanation for the different behaviour of RF molecules from the two systems. Thirdly, pulse-chase experiments with phage-infected cells suggest that RF is a precursor of RI whereas a similar type of experiment with poliovirus suggests that RF is synthesized from RI. However, the two experiments are not exactly comparable. Eukaryotes have a large pool of metabolites and in order to obtain a satisfactory chase, pulse-chase experiments are usually attempted *in vitro* rather than *in vivo*. When extracts of poliovirus infected cells are pulsed with labelled uridine and then chased, label moves from RI to single-stranded RNA and RF. Furthermore, this RF appears to be an end product since the ratio of labelled RF to labelled single strands remains constant, i.e. RF is not a precursor of single-stranded RNA (Fig. 7.4). The RI making negative single strands has never been isolated but is assumed to exist as a means to amplify the population of templates needed for virion/mRNA synthesis.

Nearly all studies on the synthesis of polio- and related viruses employ the antibiotic actinomycin D which intercalates between GC residues in DNA and prevents its transcription. Otherwise the newly synthesized cellular RNA would

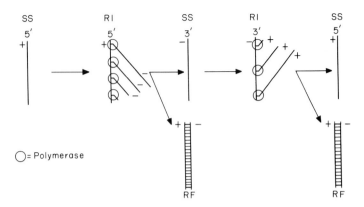

Figure 7.4 Synthesis of poliovirus RNAs. Single-stranded RNA and RF are end products.

obscure the viral RNAs. Although actinomycin D does not inhibit the multiplication of polioviruses one of the dangers inherent in using such antibiotics is that its full range of effects in the cell can never be fully appreciated. Thus what is deduced to be 'normal' viral RNA synthesis may be due to the presence of the antibiotic. In fact one of the few studies in the absence of actinomycin D showed that RF was only found when cells were treated with actinomycin D and was absent in untreated cells.

It should be apparent from the foregoing that many of the questions concerning the replication of class IV viruses remain unanswered. Rather the study of their replication has posed even more questions. Clearly this is one area of virology which requires much more investigation.

SYNTHESIS OF THE RNA OF CLASS V VIRUSES

Class V consists of viruses such as the Orthomyxoviridae, Paramyxoviridae and Rhabdoviridae which have a single-stranded RNA genome which is complementary in base sequence to the messenger RNA. Thus the synthesis of mRNA involves the transfer of information from one single-stranded RNA to its complementary RNA and requires a pre-existing RNA-dependent RNA polymerase. This enzyme is present in purified preparations of class V viruses. However, activity can only be detected after partial disruption of the virus with detergent. In the presence of the four ribonucleoside triphosphates and appropriate ions, viruses 'activated' in this way synthesize RNA *in vitro* at a linear rate for at least 2 hours. In general the *in vitro* systems yield only RNA which is complementary and sub-genomic in size and corresponds to the mRNAs made *in vivo*. Replication does not take place *in vitro* (Fig. 7.5).

When all the RNAs from an *in vitro* polymerase reaction mixture directed by a rhabdovirus, vesicular stomatitis virus (VSV), are extracted and annealed together, the virion template RNA is protected from RNase digestion and thus has been entirely copied into complementary RNA. However only a small proportion of the newly synthesized RNA is protected by the virion RNA showing

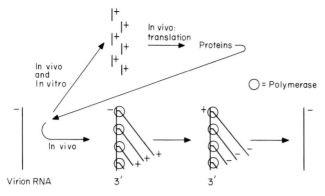

Figure 7.5 Hypothetical scheme for the synthesis of the RNAs of vesicular stomatitis virus (Rhabdoviridae).

that it is synthesized in excess of the template. In other words re-initiation of RNA synthesis takes place *in vitro*. The fidelity of RNA synthesis *in vitro* is elegantly shown by combining it with an *in vitro* translation system. Analysis by poly-acrylamide gel electrophoresis and peptide mapping demonstrates that all the viral proteins are synthesized from the novel mRNAs.

There is a unique situation which pertains to the synthesis of influenza virus RNA. Although the *in vitro* activity of influenza virus polymerase is not affected by actinomycin D or α-amanitin (the latter specifically inactivates eukaryotic RNA polymerase II which synthesizes cellular mRNA), viral RNA synthesis *in vivo* is totally inhibited if the drugs are administered early in infection. This means that *in vivo* RNA synthesis requires a cell-coded function, but its precise nature is unknown. DNA or protein synthesis is not required and no newly synthesized cellular RNA has been detected. Recently the situation became more extraordinary when influenza virions used in an *in vitro* coupled transcription-translation system were found to direct the synthesis of viral proteins even in the presence of inhibitors. Why does the drug inhibit *in vitro* but apparently not *in vivo?* Unlike the rhabdoviruses, influenza viruses do not re-initiate transcription *in vitro,* so we assume that the cellular function needed for viral multiplication has to do with re-initiation of viral RNA synthesis. There is other evidence to suggest that this particular virus-cell interaction takes place within the nucleus. Since the orthomyxoviruses are the only 'true' RNA viruses whose multiplication depends absolutely on the nucleus there may be some molecular surprises in store when the mechanisms of RNA synthesis are fully elucidated.

SYNTHESIS OF THE RNA OF CLASS III VIRUSES

Viruses such as reovirus of man and wound tumour virus of plants contain double-stranded RNA. By analogy with DNA replication this double-stranded RNA could replicate by a semi-conservative mechanism such that the comple-mentary strands of the parental RNA duplex are displaced into separate progeny genomes. Alternatively, the parental duplex could be conserved or degraded.

103

When reovirus is treated with chymotrypsin the outer capsid shell is removed and 'cores' are produced which possess RNA polymerase activity. These 'cores' are infectious and have a density in CsCl (1.45 g/cc) which is distinct from that of mature virions (1.37 g/cc) and these properties were used to distinguish between the possible modes of replication. Cells were infected with 'cores' which had previously been labelled with ^{3}H-uridine. About 15 hours after infection the cells were collected and a cytoplasmic extract prepared and analysed by equilibrium centrifugation in CsCl. All the ^{3}H-labelled parental RNA remained with the subviral particles at its original density of 1.45 g/cc while the progeny virus, as measured by plaque-forming activity, was unlabelled and had a density of 1.37 g/cc. Thus, the parental RNA is conserved during replication. How, then, is the double-stranded RNA molecule replicated?

The RNA polymerase contained in 'cores' transcribes one strand of each of the ten double-stranded segments of the reovirus genome (cf. page 56). The RNA products are single-stranded and do not anneal and since the same RNA molecules appear as messengers on the polyribosomes they have been designated as + strands. Consequently, the complementary single-strand has been designated as the − strand. These − strands are found exclusively in double-stranded RNA and do not appear free in the cytoplasm suggesting that the synthesis of the two complementary strands is asymmetric. Replication takes place in sub-viral particles which contain single-stranded + RNA and have replicase activity. These particles have been isolated and the change from single to double-stranded RNA is observed *in vitro* (Fig. 7.6).

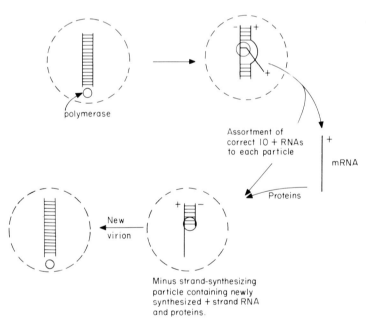

Figure 7.6 Scheme for the synthesis of reovirus RNA. Virions contain 10 double-stranded RNA molecules but only 1 is shown.

As double-stranded RNA is formed by the synthesis of a complementary strand upon a preformed single-strand, then the double-stranded RNA that is formed on pulsing the cells with ^{3}H-uridine should be preferentially labelled in one of the two strands. If both strands of the newly-formed duplex molecules were synthesized simultaneously then they should be labelled equally, regardless of the length of the labelling period. To distinguish between these possibilities L-cells were infected with reovirus and at various times the culture was pulsed with ^{3}H-uridine. The double-stranded RNA was extracted, denatured, and then re-annealed in the presence of excess unlabelled + strands. In cultures that were continuously labelled for 9 hours or more after infection, the amount of label in the + and − strands of the double-stranded RNA was the same, as indicated by the fact that annealing of the denatured double-stranded RNA with excess unlabelled + strands resulted in an annealed product containing 50% of the label present in the original double-stranded RNA. By contrast, the distribution of label in double-stranded RNA isolated from cells that had been pulse-labelled for 30 minutes, at a time when double-stranded RNA synthesis was already under way, was asymmetric with regard to the complementary strands. Over 95% of the radioactivity in the pulse-labelled double-stranded RNA was conserved in a double-stranded form after re-annealing with excess unlabelled + strands.

The assembly of the 10 segments of RNA to form a complete genome is a remarkable phenomenon. The observed high rate of generation of 'recombinants' must occur through the random assortment of segments in a common pool, presumably at the stage before free + strand RNA enters the − strand-synthesizing particle. However, the mechanism of genome assembly is entirely unexplained.

COMPARISON OF DNA AND RNA SYNTHESIS

From the foregoing discussion it is evident that the modes of replication of double-stranded DNA and RNA are unexpectedly quite different whereas those of single-stranded RNA and DNA are similar.

DEFECTIVE RNAS

Just as viruses are parasitic on cells, so defective viruses are dependent upon some function(s) provided by another virus. Even viruses have parasites! Defective-interfering (DI) and satellite viruses fall into this category and are discussed below.

The generation of defective-interfering (DI) virus RNA

All RNA and DNA viruses produce DI particles as the result of an error in their nucleic acid synthesis. This section is restricted to the RNA viruses about which more is known. DI viruses are deletion mutants which are unable to reproduce themselves without the assistance of the infectious parental virus (i.e. they are defective). For this reason propagation of DI virus is optimal at a high MOI when all cells contain a complete viral genome. DI virus depresses (or interferes with) the yield of infectious progeny as it competes for a limited amount of some product synthesized only by the infectious parent. Some biological implications of DI particles are discussed in Chapter 11.

DI RNA molecules are much shorter than the complete genome. One type of DI RNA (Semliki Forest virus) retains both the 5′ and 3′ terminal regions while another (vesicular stomatitis virus) lacks the 3′ terminus. A clue leading to one hypothesis explaining how DI RNAs are generated came from electron microscopic examination of genomic and DI RNAs from single-stranded RNA viruses. Both were found to be circularized by hydrogen bonding between short complementary sequences at the termini, forming structures called 'pan-handles' or 'stems' (Fig. 7.7A). The deletion which results in DI RNA may arise when a polymerase molecule detaches from the template and reattaches at a different point or to the nascent strand (Fig. 7.7B). Thus the polymerase begins faithfully but fails to copy the entire genome.

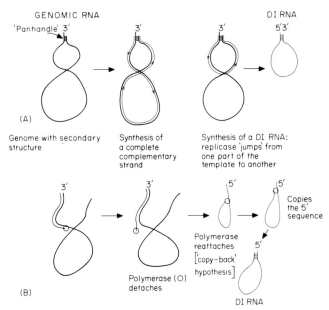

Figure 7.7 Hypothetical schemes to explain the generation of DI RNAs having sequences identical with both the 5′ and 3′ regions of the genome (A) and with the 5′ region only (B).

When DI RNAs lack the sequence coding for the transcriptase or replicase, interference results from the difference in size between the defective and complete genomes. In a given amount of time an enzyme will be able to make more copies of the smaller DI RNA. Thus as time progresses interference will become greater as the concentration of DI RNAs increases in relation to the genomic RNA. Astute readers will have recognized the interesting parallel of the generation of DI RNA with the *in vitro* evolution of the small $Q\beta$ RNAs discussed above.

Because DI viruses depend upon parental virus to provide those proteins for

106

which it does not code, DI and parental viruses are composed of identical constituents apart from their RNAs. Thus it is usually difficult to separate one from the other. A notable exception is the DI particle of the bullet-shaped rhabdovirus, VSV, which is a truncated version of the infectious particle. Consequently DI VSV became known as T particles. Other viruses can be separated from their DI particles by careful equilibrium density centrifugation which exploits slight differences in their densities. Strangely, DI Semliki Forest virus is more dense than infectious virus even though it contains about 25% of genomic RNA. Presumably DI SFV packages 5 or more molecules of DI RNA into each virion.

RNA SATELLITE VIRUSES

We have already mentioned satellite DNA viruses. There are also satellite RNA viruses, most of which occur in plants and the best-studied example is tobacco necrosis satellite virus. Satellite viruses cannot multiply without a helper virus which is specific for a particular satellite virus. Satellite viruses are distinguished from DI viruses as their RNA has little or no homology with that of the helper. Nobody knows what the origin of satellite viruses might be.

Preparations of the Rothamsted strain of tobacco necrosis virus (TNV) contain particles of two sizes. Only the larger particles are infective and they are serologically unrelated to the smaller satellite particles. The larger particles (TNV) cause characteristic necrotic lesions when inoculated on tobacco leaves; isolated satellite particles do not, nor do leaves inoculated with satellite virus yield particles. Multiplication of satellite tobacco necrosis virus reduces the size and number of lesions caused by tobacco necrosis virus and this phenomenon is used to assay the satellite virus. The satellite virus cannot replicate by itself and as a result of the interaction of the two viruses during replication, the helper virus is inhibited.

The reason why satellite TNV depends on TNV can be obtained from consideration of the smaller particle's structure. It contains a single RNA molecule 1200 nucleotides long. This corresponds to a protein 400 amino acids long which is approximately the size of a satellite virus coat protein. Thus the helper virus is required to provide the enzymes necessary for replication of the RNA of the satellite virus. The RNA of the satellite of tomato black ring virus (TBRV) can be translated faithfully *in vitro* yielding a single protein probably identical with a protein of the same size found *in vivo*. This protein represents the full coding capacity of the satellite RNA but is *not* the coat protein and has no known function. TBRV satellite has the same coat protein as TBRV itself.

Diversity among the satellite viruses of plants is exemplified by their presence in several different groups including the tobacco necrosis viruses, nepoviruses (TBRV) and cucomoviruses (cucumber mosaic virus). Furthermore unlike TNV, other satellites (such as TBRV satellite) may have no effect upon the disease caused by the helper virus. In general satellite viruses can be viewed as agents which are capable of regulating expression of a disease.

For a long time virologists have asked why it is that infectious nucleic acid can be extracted from some viruses but not from others. An examination of the Baltimore classification provides the answer. Viruses belonging to classes II, V and VI, as well as some belonging to other classes, require the transcription of their genetic information to another nucleic acid upon infection of a susceptible cell. In most cases this transfer is mediated by an enzyme associated with the virus particle such as the DNA-dependent RNA polymerase of vaccinia virus, the RNA-dependent RNA polymerase of vesicular stomatitis virus (page 102) and reovirus (page 103) and the RNA-dependent DNA polymerase of tumour viruses (page 200). Purified nucleic acid from such viruses will lack these enzymes and consequently will not be able to initiate a productive infection. Also cells do not possess an enzyme capable of using the viral RNA as template.

However, if susceptible cells are first infected with a mutant of the same virus such that only the early steps in infection occur, purified nucleic acid might be infectious since the cells will contain the required enzyme.

Bacteriophage ϕX174 belongs to class II indicating that viral messenger RNA can only be synthesized after formation of the complementary strand. Thus it might be *expected* that purified ϕX174 DNA would not be infectious. Since the viral DNA *is* infectious we can conclude that the first step in infection, the synthesis of the complementary strand, is carried out by a host cell enzyme. Clearly, the possession of infectious nucleic acid should not be the sole criterion for the assignment of any virus to a particular group.

MODIFICATIONS OF RNAs

Described below are a number of interesting properties of viral RNA molecules. Functions or biological significance of these structures are at best poorly understood.

Modification of the 3′ terminus of viral RNA

Polyadenylation

Many cellular mRNA species isolated from eukaryotic, but not prokaryotic, cells have a homopolymer of about 100 adenylic acid residues covalently bound to their 3′ terminus. This was first discovered in RNA from poliovirus virions and is now known to be present on the mRNAs of most viruses. Such RNAs are readily isolated by affinity chromatography on a column which contains poly (dT) linked to a support matrix. At relatively high ion concentration (400 mM NaCl) RNA with a poly (A) tail anneals to poly (dT) while other RNA molecules pass through. By reducing the salt concentration to 1mM, the hydrogen bonding breaks down and the poly (A)-containing RNA can be collected.

There is some uncertainty as to how poly (A) sequences are added to viral RNAs. In poliovirus there is evidence that it is added by transcription from a poly (U) tract on the complementary RNA strand. However this may not be general as others failed to find poly (U) in the RNA of another picornavirus. For the latter and probably most cellular mRNAs, poly (A) is synthesized post-transcriptively from the addition of ATP molecules. This view is supported by the finding of polyadenylating activity in purified Newcastle disease virus.

The role of poly (A) is not clear. Poliovirus RNA from which poly (A) has been removed by enzymatic digestion has reduced infectivity but Q*β*and TMV RNAs are not polyadenylated and are infectious. In *in vitro* systems, poly (A) does not affect translational efficiency of the message but confers stability to the message both *in vitro* and *in vivo*. However reovirus mRNAs are entirely lacking in poly (A). In general, the length of the poly (A) is characteristic of individual viruses: poliovirus has a tract of about 80 and another picornavirus, EMC, about 14 adenylic acid residues. Thus we are almost driven to the conclusion that the sole function of poly (A) tracts is to help virologists to isolate polyadenylated RNAs! Hopefully the real significance of terminal poly (A) sequences will soon be discovered.

Binding of amino acids by plant virus RNAs

The RNAs of a number of plant viruses have been studied and are not poly-adenylated. However the 3′ end can fold to form a structure resembling tRNA which can accept amino acids. The significance of this is not known and there is no evidence that the viral RNA functions as a tRNA in the cell. In *in vitro* protein synthesis the amino acid-viral RNA complex substitutes poorly for tRNA. Each viral RNA has specificity for one amino acid, e.g. tobacco mosaic virus and turnip yellow mosaic virus bind histidine and valine respectively.

Modifications of the 5′ terminus of viral RNA

Capping

Most eukaryotic mRNAs have their 5′ terminus modified to form a cap structure (see also p. 55). This is made contrary to the rules of nucleic acid synthesis (Fig. 7.8): firstly the penultimate G of the cap is added to the 5′ terminus of the message proper and secondly the terminal G is joined by a 5′-5′ linkage. An additional point to notice is that both Gs are methylated. This is a modification common to cellular mRNAs and is of unknown significance. The cap is synthesized by a series of 5 enzymes normally shortly after transcription has started. In reovirus mRNA, the cap is necessary for the formation of a stable initiation complex for protein synthesis but the requirement is not absolute in other mRNAs. Some mRNAs are not capped with m^7G but are translated *in vitro*.

$$ppp\ G\ +\ ppp\ N.pN \rightarrow ppp\ G.pN.pN\ +\ 2\ Pi$$
$$ppp\ G\ +\ ppp\ G.pN.pN \rightarrow G\ ppp\ G.pN.pN\ +\ 3\ Pi$$
$$2\ \text{S-adenosylmethionine}\ +\ G\ ppp\ G.pN.pN \rightarrow m^7\ G\ ppp\ G^m.pN.pN\ +\ 2\ \text{S-adenosylhomocysteine}$$

Figure 7.8 Reactions involved in the synthesis of the cap of reovirus mRNA. 'Deleted' phosphate groups are those lost during cap formation.

Covalently linked protein

Poliovirus virion RNA is not capped but has a small protein (mol. wt. about 7000) covalently linked to its 5′ terminus. The protein is absent from polysomal + strand RNA, thus it seems that it has to be removed before the virion RNA can be translated. Nascent + strands on the replicative intermediate also have the protein attached, which may indicate that its removal is the process which decides if newly synthesized + strand RNA ends up as message or in progeny virions. This interesting situation has a parallel with the adenovirus genome which is circularized by a protein linked covalently to each 5′ end of the double-stranded DNA (p. 53).

8 The process of infection: III the regulation of gene expression

When a virus infects a cell not all the genes are expressed at the same time and this was suggested in Chapter 1 when we discussed the pulse-chase experiments of Hershey and co-workers. These workers noted the existence of 'early' and 'late' proteins but more definitive proof of regulation comes from a systematic study of the appearance of phage T4 proteins. Infected cells were pulsed with radioactive leucine, the proteins extracted, electrophorezed on acrylamide gels and the protein levels estimated by autoradiography. These studies showed that not all proteins are synthesized at the same time nor are many synthesized continuously. Rather, some are synthesized for only a few minutes early in infection, others are synthesized late in infection while others are synthesized for about half the infection cycle. Most of our knowledge comes from studies on bacteriophage regulation but there is a growing body of information on regulation in eukaryotic cells and the mechanisms regulating the development of animal viruses are considered in the second part of this chapter. There is still little knowledge about regulation and gene expression in plant viruses.

DNA BACTERIOPHAGES

During gene expression, genetic information in DNA nucleotide sequences is transcribed into complementary RNA sequences and these are translated into the polypeptide chains which form the ultimate products of the gene. Control of gene expression might occur at the level of transcription or translation but it is becoming increasingly clear that in bacteriophage-infected cells transcriptional control is of major importance.

Transcription of DNA into RNA is carried out by the enzyme RNA polymerase

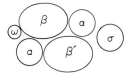

Figure 8.1 Structure of RNA polymerase from *E. coli.*

which, in *E. coli* at least, has a rather complicated structure. The enzyme from *E. coli* consists of five different polypeptide chains—$\alpha, \beta, \beta', \omega$ and σ held together by hydrogen bonding in the ratio $\alpha_2\beta\beta'\omega\sigma$ (Fig. 8.1). The polypeptide chain σ, or sigma factor, can easily be separated from RNA polymerase and the rest of the structure, the core enzyme, retains its catalytic activity. Thus the sigma factor does not contribute towards the catalytic activity of the enzyme but it still has an important role to play in RNA synthesis for in its absence correct initiation of RNA synthesis fails to occur *in vitro*. Thus, the most likely role of the sigma factor is recognition of start signals on the DNA molecule.

In Chapter 1 we noted the existence of 'early' and 'late' proteins during the infection. Thus at late times after virus infection new sets of genes are transcribed which were not transcribed early in infection. Based on our knowledge of transcription, how might this occur? There are at least four possible models:

1.   The initiation specificity of the host RNA polymerase might be altered by synthesis of a new sigma factor or by modification of the core enzyme.
2.   There could be *de novo* synthesis of a new polymerase having a new initiation specificity.
3.   There could be synthesis of a protein factor which antagonizes normal termination of RNA chains (anti-termination factor) and hence allows host RNA polymerase to read through to distal genes.
4.   There could be synthesis of a positive control protein which is required for initiation of RNA synthesis at certain promoter sites but which is not a component of RNA polymerase.

These mechanisms are not necessarily exclusive and in the case of bacteriophage T7 and T4 more than one of them is utilized. Since the pattern of T7 transcription is the simplest we shall discuss it first.

The regulation of bacteriophage T7 development

During T7 infection a number of discrete mRNA molecules are formed and since these are relatively stable they can be isolated and characterized. These mRNA molecules can be divided into two classes: early mRNA which consists of 4 species and late mRNA consisting of 8 or 9 discrete species. The two complementary strands of T7 DNA can be separated by one of the methods described in Chapter 3, and it is found that both early and late mRNA synthesized *in vivo* anneals to only one of the two strands. Thus only one DNA strand is used as a template for mRNA synthesis.

Only early mRNA is made during infection with gene 1 conditional lethal mutants. Thus the gene 1 product is necessary for transcription of late mRNA. Initially it was suggested that the gene 1 protein was a σ-like factor which alters the initiation specificity of the bacterial RNA polymerase after infection. However, it has now been shown that the gene 1 product is in fact a new T7 specific RNA polymerase. The purified enzyme consists of a single protein component when analysed by SDS gel electrophoresis. Both the protein and the enzyme sediment at about 7S suggesting that the active enzyme contains only a single polypeptide chain and consequently it is much simpler than the corresponding host enzyme. Furthermore, the T7 enzyme is resistant to rifamycin, an antibiotic inhibiting the host RNA polymerase, and to antibody directed against the host enzyme.

The current picture of T7 development involves transcription of early genes by the host cell RNA polymerase. Termination occurs at a specific site located between genes 1.3 and 1.7 and without the intervention of the termination factor ρ. *In vitro, E. coli* RNA polymerase synthesizes a single mRNA species equal to the entire early region in length. However, *in vivo* this mRNA is cut into discrete monocistronic messengers by the host enzyme RNase III. Transcription of gene 1, an early gene, gives rise to a new RNA polymerase which transcribes late T7 genes, i.e. those genes which are inaccessible to the bacterial RNA polymerase. A number of discrete classes of late T7 mRNA can be isolated from infected cells. Each class is initiated at a different promoter site for T7 RNA polymerase but all terminate at a site close to the right-hand end of the T7 DNA molecule.

The question remains as to how T7 turns off early phage and bacterial functions. The product of a late gene is required for the turn off of host RNA synthesis and since this protein is able to bind to bacterial RNA polymerase it is possible that this enzyme is inactivated thus preventing further synthesis of either host mRNA or early T7 mRNA.

The pattern of T7 regulation just described correlates nicely with genetic studies on T7. With a mol. wt. of 25×10^6 daltons, T7 has a coding potential of approximately 30 genes. So far, 25 genes have been identified, mostly with the aid of amber mutants, and these can be ordered on a linear genetic map (Fig. 8.2)

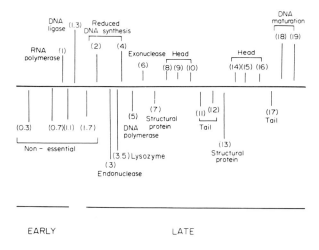

Figure 8.2 The genetic map of bacteriophage T7. The numbers in parentheses are the gene numbers. Note the clustering of early and late genes.

by means of standard genetic crosses and deletion mapping. The T7 specified proteins in infected cells can be analysed by electrophoresis on acrylamide gels. Any protein affected by an amber mutation or a deletion should disappear from its normal position in the gel and this permits the identification of each band as the product of a particular gene. Studies on the time course of synthesis of different gene products shows that the proteins specified by genes 0.3 to 1.3 are synthesized early while all the other genes are synthesized late. Thus all the early genes are clustered together at the end of the genetic map while the late genes form a larger cluster. Furthermore, since mRNA is synthesized in a 5′ to 3′

113

direction, the early genes must be clustered at the 3′ end of the transcribed strand.

Late in infection there should be sufficient amounts of the enzymes involved in DNA replication that their continued synthesis would be unnecessary. In this respect it is interesting to note that the genes determining the synthesis of these enzymes are not expressed throughout infection but are shut off sometime before lysis occurs. However, the mechanism controlling this is not known and no mutants are known in which this shut-off does not occur.

The regulation of bacteriophage T4 development

The regulation of transcription in cells infected with phages T4 and λ is far more complex than that of T7. We shall postpone our discussion on regulation in λ until Chapter 10, because it is intimately tied in with the phenomenon of lysogeny, and consider only T4 regulation here. At least three classes of mRNA are found in T4 infected cells: immediate early, delayed early and late mRNA. As the name implies, immediate early mRNA is the first viral specified RNA synthesized in the cell. Delayed early RNA makes its appearance about 2-3 minutes after infection and late RNA is synthesized from about 10 minutes onwards. Furthermore, in contrast to T7, all T4 specific RNA synthesis is sensitive to rifamycin throughout the developmental cycle. Since the β sub-unit of the host polymerase is the site of rifamycin sensitivity, this sub-unit must be necessary for all phage RNA synthesis. Furthermore, pre-labelling of RNA polymerase before infection has shown that the α and β′ sub-units are also conserved although they are modified slightly. By two minutes after infection, when delayed early RNA synthesis begins, 5′ adenylate is covalently added to the α sub-unit. Ten to fifteen minutes after infection, when late RNA synthesis starts, the mobility of the β′ sub-unit in electrophoretic gels is observed to be different suggesting that it too is modified. Furthermore, a newly-synthesized polypeptide has been isolated from the core polymerase early in infection which could be a phage-specified ω factor. Thus, at different times, three different core polymerases are present in the infected cell (Fig. 8.3), the last two to appear being modifications of the host polymerase.

Delayed early cistrons are contiguous with the immediate early cistrons and

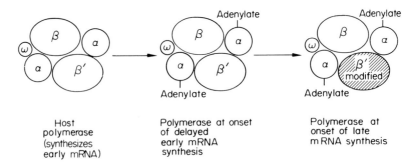

Figure 8.3 Modification of the *E. coli* RNA polymerase following infection with bacteriophage T4.

both immediate early and delayed early mRNA are transcribed from the same DNA strand. To explain the synthesis of delayed early RNA two models have been proposed. The first model suggests that delayed early transcription results from the inhibition of ρ-mediated chain termination thus allowing polymerases which had been initiated at immediate early promoters to read through into delayed early regions. This process could occur *in vivo* because early RNA sequences can be found in long RNA molecules which were initiated immediately after infection when only immediate early promoters are active. The second model suggests that delayed early RNA is synthesized as a consequence of the altered initiation specificity of RNA polymerase itself. A factor has been isolated from T4 infected cells which directs both host and 5′ adenylated RNA polymerase to transcribe delayed early RNA together with some immediate early RNA. In similar conditions the *E. coli* σ factor directs the synthesis of 80-100% immediate early RNA. These results suggest that the early modifications to the core polymerase do not influence the accuracy of promoter selection.

Late RNA synthesis is absolutely dependent on the product of T4 gene 55. When the RNA polymerase DNA complex is isolated from cells infected with gene 55 mutants, late RNA synthesis could be induced by addition of a crude preparation of gene 55 product. Also a factor has been isolated from cells late in the T4 infection process which stimulates the late RNA polymerase, but not the host polymerase, to preferentially transcribe late RNA. However, as yet there is no evidence linking this factor with the gene 55 product nor is its mode of action known.

In the absence of T4 DNA replication late RNA is not synthesized unless DNA ligase is inactivated. Binding studies have shown that the late polymerase binds preferentially to denatured DNA and it is possible to interpret these two results as follows. A specificity factor, possibly the gene 55 product, could either act as a σ factor enabling the late polymerase to recognize certain breaks in lieu of promoters or could itself bind to the breaks and generate a receptive structure for core polymerase.

Whereas T7 utilizes only one of the possible mechanisms of regulation listed on page 112, namely the formation of a new RNA polymerase, T4 uses at least three other mechanisms. Thus, studies on T4 development are more complex and from the foregoing it is clear that much has still to be learned.

RNA BACTERIOPHAGES

The genomes of the small RNA bacteriophages such as MS2 and R17 only code for three proteins: phage coat protein, maturation protein (A protein) and the enzyme RNA synthetase. It might be thought that with such limited genetic potential these phages can reproduce without the use of any control mechanisms. This is not so. If susceptible cells are infected with these phages in the presence of radioactive amino acids and then extracted, the amount of radioactivity in each of the phage proteins can be measured after electrophoretic separation. From experiments of this type it is clear that the coat protein is synthesized in great excess over the other two proteins. In fact, the three proteins are produced in the ratio of 20 coat proteins: 5 synthetase molecules: 1 maturation protein.

Regulation by means of a polarity gradient

Initially it was suggested that the order of the genes from the 5′ end to the 3′ end of the molecule was coat protein—synthetase—maturation protein. If ribosomes could only attach at the 5′ end of the messenger, then regulation could be achieved if some of the ribosomes dissociated from the messenger at the end of each gene so that decreasing amounts of each protein are produced. Regulation would thus depend on the existence of a polarity gradient. This idea is consistent with the finding that translation of the coat protein gene is necessary before it is possible to translate the synthetase gene since nonsense mutations in the coat protein gene prevent expression of the synthetase gene.

Unfortunately the polarity model of regulation is unsatisfactory since it does not explain the kinetics of synthesis of the different proteins. Examination of infected cells shows that coat protein synthesis is exponential while the A protein is synthesized at a rate which is very nearly linear. The replicase, in contrast, is synthesized exponentially early in infection but synthesis is turned off late in infection. Since the RNA of these phages does not undergo recombination it has not been possible to confirm the proposed order by conventional genetic analysis but a genetic map has been produced by biochemical means. Each of the three phage genes commences with an initiating sequence of nucleotides which can bind ribosomes directly and if a complex of ribosomes and phage RNA is digested with nucleases, these initiating regions are shielded against enzymatic attack. These regions can thus be isolated and sequenced. The enzyme ribonuclease IV from *E. coli* cleaves R17 RNA at a position 40% along the molecule from the 5′ end. The initiating regions present on each of the two ribonuclease IV fragments can also be identified following nuclease degradation of fragment-ribosome complexes. In this way the initiation sequence for the maturation protein and the coat protein were shown to be on the 40% fragment. Furthermore, nucleotide fingerprints corresponding to the internal parts of the gene appear in the 60% fragment. These data unambiguously show that the order of the genes in the phage must be:

5′—maturation protein—coat protein—synthetase—3′

This gene order rules out polarity as a method of controlling development. How then is the differential expression of the three genes to be accounted for?

Translational control of development

It now appears that differences in the frequency of initiation of synthesis of the three proteins is the basis for regulation of RNA phage development. However, before presenting the evidence on which this statement is based let us consider briefly how initiation of synthesis of each protein can be assayed. The simplest assay determines the amount of each initial dipeptide formed in an *in vitro* protein synthesizing system under conditions where chain elongation is inhibited. Since

116

the N-terminal amino acid of each phage protein is different, the initial dipeptides are distinctive. In the case of MS2 and R17 they are fmet-ala for coat protein, fmet-ser for synthetase and fmet-arg for the maturation protein. The basis of the assay is to determine the amount of radioactivity incorporated into each fmet-labelled dipeptide following separation by paper electrophoresis. Another initiation assay depends on the ability of ribosomes in an initiation complex to protect against ribonuclease digestion that region of the phage RNA which is bound to the ribosome.

The effect of RNA conformation

The relative amounts of coat protein, synthetase and maturation protein initiated in *E. coli* extracts depend on the integrity and conformation of the viral RNA. With native RNA isolated from phage particles, only coat protein is initiated, as shown by the initial dipeptide assay. However, alteration of the conformation of the viral RNA by formaldehyde treatment, or fragmentation of the RNA, allows all three proteins to be initiated. The similar effects of these two treatments strongly suggest that specific conformational features in native RNA restrict initiation at the other two sites.

Recently, determination of the nucleotide sequence of an extensive stretch of the coat protein and adjacent synthetase cistrons has revealed a probable region of hydrogen bonding, 21 nucleotides long, between codons 24 to 32 of the coat cistron and the synthetase initiation site. Hydrogen bonding in this region could account for the inability of ribosomes to bind to the synthetase initiation site. Furthermore, an amber mutation in any of the first 24 codons of the coat cistron would prevent ribosomes opening up this double-stranded region, thus accounting for the polar effect of some coat protein mutants on initiation of synthetase.

In contrast with synthetase formation, initiation of the maturation protein is not affected by conformational changes that occur during translation of the coat gene. Thus, amber mutations in the coat cistron have no polar effect on production of maturation protein *in vitro*. How then is the initiation of the synthetase regulated? As indicated in Chapter 7, much of the RNA in the infected cell exists as replicative intermediate (RI) and it will be recalled that this consists of an intact − strand template with one or more nascent single-stranded plus chains extending from it. *In vitro,* RI initiates about five times as much maturation protein, relative to the total protein initiated, as does single-stranded RNA isolated from phage particles. Presumably the conformation of the 5′ end of the single-stranded RNA chains in the RI is different from that of completed RNA molecules, allowing ribosomes access to the initiation site for the maturation protein on the nascent strands. Only a fraction of the maturation protein molecules initiated on the RI *in vitro* are actually completed, in contrast with coat protein which is the major product directed by RI. This failure to produce complete molecules of maturation protein could be explained by assuming that only the shortest nascent strands have open initiation sites for the maturation protein, and that these strands are not long enough to contain the entire cistron. According to this hypothesis, as each nascent RNA chain is synthesized in the infected cell, a ribosome attaches to the maturation protein initiation site in the short 5' tail; as the RNA chain is elongated, that ribosome proceeds to translate the maturation protein cistron, but folding of the elongated RNA strand (Fig. 8.4) prevents additional ribosomes from initiating at the initiation site.

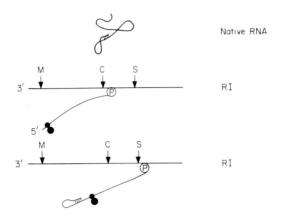

Figure 8.4   Model to account for the linear rate of production of maturation protein. In native RNA the initiation site for maturation protein synthesis (M) is not available to the ribosome because of hydrogen bonding. During replication an RNA tail is produced at the 5' end in which the initiation site is available to the ribosome. As the nascent RNA chain grows in length, hydrogen bonding again prevents ribosome binding. P = polymerase, C = coat protein, S = synthetase.

Other factors affecting translational control

It appears that conformation of the phage RNA is not the only factor influencing the independent translation of the three genes for there is some evidence indicating that the specificity of the ribosomes and initiation factors are also important. Since the ribosome binding sites differ in nucleotide sequence for each phage cistron it is possible to imagine that the same ribosome can recognize all three initiation sites but bind with different affinity to each. Alternatively, a separate class of ribosomes might recognize the initiation sequence of each cistron.

The strongest evidence that ribosomes can discriminate among cistrons is the observation that whereas *E. coli* ribosomes initiate the coat protein when primed with native phage RNA, *Bacillus stearothermophilus* ribosomes initiate only maturation protein. If the 30S sub-units from *B. stearothermophilus* ribosomes are mixed with the 50 S sub-units from *E. coli* ribosomes, and the hybrid ribosomes primed with native phage RNA, only the maturation protein is initiated. Also, if the alternative hybrid is used, only the coat protein is initiated. Thus the cistron specificity is a function of the 30 S sub-unit. These findings raise the interesting possibility that within infected *E. coli* cells, a different class of ribosomes might be responsible for initiating each cistron. This suggestion is not too improbable since the 30 S sub-units are known to be structurally heterogeneous with respect to certain proteins. Furthermore, when MS2 infected *E. coli* are exposed to kasugamycin, an antibiotic which acts on the 30 S sub-unit to inhibit initiation, synthesis of maturation protein is inhibited relative to coat protein synthesis. This

118

observation suggests that either ribosomes of different kasugamycin sensitivity initiate translation of the maturation and coat protein cistrons or that the antibiotic differentially affects the ability of a single population of ribosomes to bind to the two initiation sites of the phage RNA.

Formation of initiation complexes with coliphage mRNA requires 30 S ribosomal sub-units, GTP and several initiation factors before the 50 S sub-unit is added to yield an active 70 S ribosomal complex. The initiation factors have been fractionated into three proteins designated IF1, IF2 and IF3 and of these IF3 is particularly important because it is required for stable binding of the 70 S ribosomes to the initiation sites. There is now some evidence to suggest that multiple species of IF3, with cistron selective activities, are present in *E. coli.* When a mixture of ribosomes and initiation factors from T4 infected *E. coli* are primed with MS2 RNA, analysis of the oligonucleotide sequences protected by bound ribosomes reveals that appreciable binding only occurs at the initiation site for the maturation protein. The deficiency in the T4 initiation factor preparation has been specifically identified as IF3 suggesting that there is a selective inactivation of coat and synthetase-specific initiation factors following T4 infection. Thus uninfected *E. coli* may contain components of IF3 which discriminate between these cistrons and that of the maturation protein.

The three methods of regulation of RNA phage development which have just been described do not account for the cessation of synthesis of the RNA synthetase late in infection. This observation can be accounted for by a fourth mechanism of translational control—the specific binding of phage proteins to the messenger. When *E. coli* is infected with phage carrying mutations in the coat protein, there is an enhancement of synthetase formation suggesting that coat protein acts as a translational repressor. Furthermore when coat protein is incubated with phage RNA, and the RNA then used as a messenger *in vitro,* the formation of synthetase is specifically inhibited as shown by the dipeptide assay. By way of contrast, the presence of coat protein does not affect the initiation or elongation of either the coat protein or the maturation protein. Thus the role of coat protein as a translational repressor of synthetase formation has been established.

ANIMAL VIRUSES

The regulation of gene expression by animal viruses can occur at the level of transcription or translation (or both) and to a degree which varies not only between different viruses but also during a multiplication cycle. The bewildering permutation of events which can ensue is further complicated by the possible sequestration of events in the virus multiplication cycle in the cell nucleus, or an inhibition of host-cell macromolecular synthesis by viral components. Examples showing some of the variety existing are given in Table 8.1. While it is clear that many animal viruses do indeed regulate their gene expression it is not yet known in many cases, how they achieve this. However, investigation of animal viruses has led to the elucidation of mechanisms totally different from those found in bacterial systems. Thus animal viruses, as well as being of interest in their

Table 8.1 Some properties of animal virus nucleic acid and protein synthesis. (Adapted from A. E. Smith (1975) in *Society for General Microbiology Symposium* **25**, 187.)

Class	Group	Example	Genome				no. pieces mRNA	virion polymerase	nucleus involved	early/ late phases	poly A in mRNA	poly-cistronic mRNA	host shut-off
			nucleic acid	mol. wt. × 10⁻⁶	no. pieces							
I	Papova	SV40 polyoma	DNA	ds	3	1	3	–	+	+	+	+ & –	–
I	Adeno	adeno	DNA	ds	20-30	1	several	–	+	+	+	–	+
I	Herpeto	herpes simplex	DNA	ds	54-90	1	several	–	+	+	+	–	+
III	Reo	reo	RNA	ds	15	10	10	+	–	+	–	–	+
IV	Picorna	polio	RNA	ss	2.5	1	1	–	–	–	+	+	+
IV	Toga	Semliki Forest	RNA	ss	4	1	2	–	–	–	+	+	+
V	Rhabdo	VSV	RNA	ss	3.8	1	5	+	–	–	+	–	+
V	Paramyxo	NDV	RNA	ss	7	1	several	+	–	–	+	–	+ & –
V	Orthomyxo	influenza	RNA	ss	4	8	8	+	+	+	+	–	+
VI	Retro	RSV	RNA	ss	10	3-4	3	+	+	–	+	+	–

own right as infectious agents, are powerful tools for the study of the molecular biology of eukaryotic cells.

RNA viruses: Class IV: Post-translational cleavage of sub-genomic mRNA

This group includes the picornaviruses, which appear to exert no control over gene expression, and the togaviruses which show transcriptional control. The entire genome of poliovirus is translated as a huge polyprotein (mol. wt. 250 000) which is cleaved in a series of ordered steps to form smaller functional proteins (Fig. 8.5). Such post-translational cleavage is not found in bacteria but after

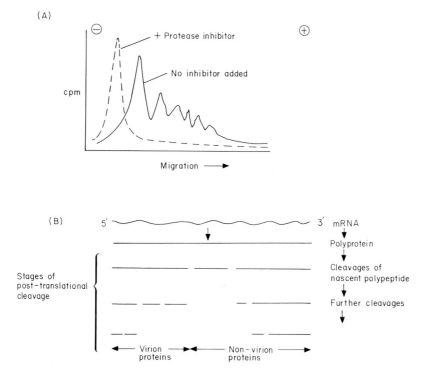

Figure 8.5 (A) Polyacrylamide gel electrophoresis of proteins synthesized in poliovirus-infected cells in the presence and absence of protease inhibitors. (B) Simplified representation of the cleavages of the polyprotein and smaller products to yield mature viral proteins (see also p. 147).

the pioneering work with poliovirus, is acknowledged as common in eukaryotic cells. Cleavage starts while the polyprotein is still being synthesized and only by inhibiting protease activity or altering the cleavage sites through the incorporation of amino acid analogues can the single polyprotein be isolated. Cleavage can also be demonstrated by pulse-chase analysis. In theory, all picornavirus proteins should be present in equimolar proportions but this is not found in practice. It seems that a differential is introduced by degradation of some of the viral proteins; for instance it is known that the virus-specified polymerase activity is unstable.

Although the Togaviridae are also Class IV viruses they have a more complex strategy for controlling protein synthesis than the picornaviruses. Togaviridae comprise two subgroups, the alphaviruses (e.g. Semliki Forest virus) and the flaviviruses. Fig. 8.6 shows a scheme for gene expression in the alphaviruses

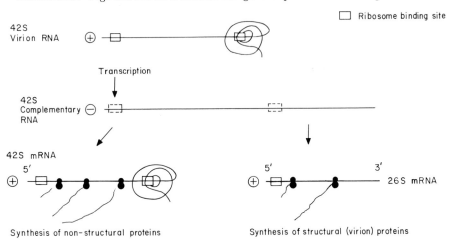

Figure 8.6 Separate control of synthesis of structural and non-structural proteins in alphavirus-infected cells is achieved by blocking of the internal ribosome-binding site in the virion RNA and the synthesis of a separate smaller mRNA.

which involves two mRNAs. However cell-free systems show that virion proteins are not made from the virion RNA which infers that it has an internal initiation site which is hidden, presumably within the secondary structure. It is not known how transcription is controlled but since no 26S RNA$^-$ † is found in the cell we can assume that virion and 26S RNA$^+$ are transcribed directly from the 42S RNA$^-$. Despite this elaborate process both mRNAs function simultaneously and the 26S RNA appears to be an adaptation for making large amounts of virion proteins without an equimolar production of non-structural elements. Like the picornaviruses, the 26S RNA and non-structural region of the alphavirus genome are translated as polyproteins. Pulse-chase experiments and tryptic peptide maps show clearly that the small functional proteins are derived from larger precursor molecules.

Yet another strategy for control of gene expression is shown by the flaviviruses, a subgroup of the togaviruses which includes yellow fever virus. Recent work has demonstrated the presence of internal initiation sites within the viral genome. This is the type of punctuation found in small RNA phages and before the work with flaviviruses was thought to exist only in prokaryotes. The flavivirus work was controlled by parallel experiments with a picornavirus and a togavirus which demonstrated their characteristic patterns of protein synthesis. However, confirmation that internal punctuation is used in eukaryotes has yet to come. The difference between flavi- and alpha-viruses is so fundamental that, if it is

† From the Baltimore scheme (Chapter 5), RNA$^+$ is the same sense as mRNA and RNA$^-$ is complementary.

122

substantiated, we can expect to see them reclassified in different virus groups.

Both picorna- and toga-viruses shut off host macromolecular synthesis, possibly so that the cells entire resources are devoted to the production of viral components, and it is thus no surprise that the cell dies. In poliovirus-infected cells, a virus-coded protein initiates the shut off; viral RNA synthesis is not required since the same degree of inhibition of cellular macro-molecular synthesis is observed when poliovirus replication is inhibited by a low concentration of guanidine hydrochloride. The effects on ribosomal RNA synthesis are comprehensive and well documented: there is inhibition of synthesis of the 45S rRNA precursor, of its processing to 28 and 18S rRNA, of their incorporation into subribosomal particles in the nucleolus and the transport of these to the cytoplasm. Cellular protein synthesis is also inhibited but little is known of the mechanism.

RNA viruses: Class V: one mRNA → one protein

These viruses, whose genomes are complementary in sequence to viral mRNA, synthesize proteins from monocistronic mRNAs. In the Orthomyxoviridae the genome consists of a number of RNA segments (usually eight) but in the Paramyxoviridae and Rhabdoviridae it is a single molecule.

Of the Orthomyxoviridae, most is known about the type A influenza viruses of man and other animals. From each genomic RNA segment a monocistronic mRNA is synthesized. The eight viral proteins are not present in equimolar amounts in the infected cell and the relative proportion of proteins changes during infection. Clearly, viral protein synthesis is controlled and the primary control operates at the level of transcription although the molecular details are not known. However the mechanism which distinguishes between mRNA$^+$ and RNA$^+$, which is template for virion RNA$^-$, is understood. Careful annealing studies have shown that the virus synthesizes two types of RNA$^+$: template RNA$^+$ is a faithful complement while mRNA$^+$ lacks some sequence from the 5$'$ end of the virion RNA (Fig. 8.7). Thus mRNAs cannot be templates for the synthesis

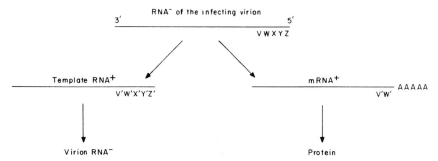

Figure 8.7 Discrimination between mRNA$^+$ and template RNA$^+$ by the synthesis of two types of complementary RNA in influenza virus type A-infected cells. Note also that only mRNA is polyadenylated and that just one of the 8 segments of the genome is shown here.

of full length genome segments. The orthomyxoviruses are unique amongst 'true' RNA viruses (i.e. excluding the retroviruses which have a DNA intermediate) in having part of their multiplication cycle within the nucleus. For example, immediately after infection the uncoated particle is transported to the nucleus while later in infection certain of the newly synthesized proteins migrate from their site of synthesis in the cytoplasm into the nucleus. Since passage of molecules across the nuclear membrane is a highly selective process, this constitutes a level of control which is unique to eukaryotic cells. The mechanism is also important to those DNA viruses which replicate in the nucleus and will be discussed later.

Of the Class V viruses with a non-segmented genome, most is known about vesicular stomatitis virus (VSV, Rhabdoviridae). From a genome of 3.8×10^6 mol. wt. five mRNAs are synthesized ranging in approximate mol. wt. from 0.3 to 1.6×10^6. Although the mRNAs are synthesized sequentially from a single initiation site, it is not known if they are derived by cleavage from a single precursor poly mRNA or if the polymerase re-initiates after finishing the synthesis of the preceding mRNA. Evidence of control of gene expression comes from two sources. Firstly, the various mRNAs are not present in the cell in equimolar proportions and, secondly, some proteins are present in greater amounts than others. As yet it is not known if this is due to relative efficiencies of translation of the different mRNAs and/or to the amount of each mRNA present. Since mRNAs are synthesized sequentially from the genome from a single initiation site, the latter could only be achieved by differential degradation for which mechanism there is, as yet, no evidence.

All Class V viruses have monocistronic mRNAs and it might be expected that post-translational cleavage would not be involved. However after its synthesis, the influenza virus haemagglutinin protein undergoes a single protease cleavage (Fig. 8.8). Uncleaved molecules occur in cells deficient in protease

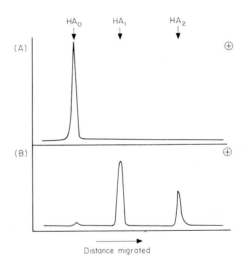

Figure 8.8 Cleavage of influenza virus haemagglutin. (A) Electrophoresis of an uncleaved molecule from non-infectious virus grown in cells deficient in protease activity. (B) Haemagglutin molecules isolated from the same virus incubated with 100 µg/ml trypsin for 30 minutes at 37°C. After incubation with trypsin the virus becomes infectious.

activity but are still incorporated into virus particles which are morphologically normal and can agglutinate red blood cells. However, such virus particles are non-infective until the HA protein is cleaved. Precisely how this step relates to infectivity is not known. Activation of cellular proteins in eukaryotes by proteolysis is well known: for instance, conversion of inactive pro-insulin to insulin and of inactive trypsinogen into trypsin.

RNA viruses: Class III: viral proteins positively control mRNA synthesis.

Control of protein synthesis in viruses with a segmented double-stranded RNA genome arises from the interplay of transcriptional and translational activities. Each genomic segment is transcribed as a monocistronic mRNA but when protein synthesis is inhibited in the cell by the addition of cycloheximide only 4 mRNAs are made, showing that there is positive control by viral proteins over the synthesis of the remaining 6 mRNAs. Further indication of transcriptional control is seen when the polysomes from an infected cell are analysed for their content of viral mRNAs. All 10 mRNAs are represented but not in equimolar amounts. In addition there is evidence of translational control since there is more protein made from mRNAs of the 0.7×10^6 mol. wt. class than those of 0.4×10^6 mol. wt., even though more of the latter is synthesized. Interestingly, none of the reovirus mRNAs are polyadenylated but they obviously still function effectively.

The 'DNA' viruses: Classes I, II and VI: using viruses to understand nuclear controls.

All DNA viruses, with the exception of the Poxviridae, synthesize their mRNA from a double-stranded DNA molecule which is located in the cell nucleus and thus they come closest as model systems for studying the synthesis of cellular mRNA. The finding that cell-coded histones, which form the major protein component of cellular DNA, are also complexed with papovavirus particle DNA underlines the similarity of these viral and cellular nucleic acids. The range of genome size spans two orders of magnitude, from parvovirus DNA of 1.5×10^6 mol. wt., to vaccinia virus DNA of 160×10^6 mol. wt., and the problem of analysing these genomes is in proportion. The reason why large viruses exist at all when the smallest function perfectly well is obscure. Certainly the larger viruses are less dependent upon the cell and can multiply when cells are not in S phase of the cell cycle, whereas the smallest viruses are unable to do so. Viruses which have a genome of intermediate size have the capacity to induce the host cell to enter S phase.

Since all except the smallest DNA viruses are capable of transforming cells (Chapter 15) one can speculate that transformation may be an abberant expression of the interaction of virus with the mechanism that regulates cell division. Much viral information appears to duplicate that present in dividing cells. However it seems likely that such functions are essential during natural infections when the virus is found in highly differentiated cells, such as neurones, which are not dividing or are metabolically deficient in some aspect.

Papovaviruses

Studies of the papovaviruses, SV (simian virus) 40 and polyoma, have been most revealing. Other virus groups will be mentioned only when their gene expression differs in a significant aspect. Papovaviruses have a double-stranded circular genome which codes for early proteins and late proteins translated from early and late mRNAs (Fig. 8.9). Papovavirus DNA is transcribed, presumably by

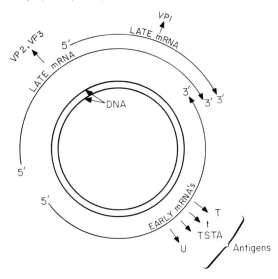

Figure 8.9 Map of the genome of SV40 virus. VP1, VP2 and VP3 are virion proteins.

cellular polymerase II which synthesizes cellular mRNAs. The early viral mRNA is synthesized from a different strand to the late mRNAs. In SV40 the major early protein is the (big) T (for tumour)-antigen which migrates to the nucleus and is thought to modify DNA so that it can be replicated. SV40 also synthesizes another tumour antigen called little t, while polyoma virus has three different tumour antigens. Their roles in transformation are discussed later, on p. 193. Other SV40 early proteins, but of unknown function, are the U-antigen, which is also in the nucleus, and the TSTA (tumour-specific transplantation antigen) in the plasma membrane. The early phase is followed by induction of the synthesis of host cell enzymes and host cell DNA and is independent of these cellular processes. Viral DNA synthesis ensues, followed by viral late mRNA. Late proteins are then assembled with DNA into progeny virions. Early and late mRNAs are transcribed from different strands of DNA. However, strand selection does not appear to represent a control process since both early and late mRNAs are both found in the nucleus late in infection. The predominant synthesis of late proteins is apparently achieved by allowing only the transport of late mRNA transcripts across the nuclear membrane into the cytoplasm (Fig. 8.10). It is interesting that the smaller late-mRNA is a subset of the larger (Fig. 8.9), a situation parallel with the virion and 26S mRNA of alphaviruses (Fig. 8.6).

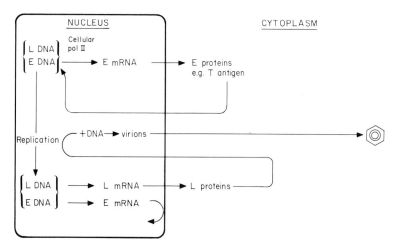

L = Late

E = Early

Figure 8.10 Summary of the major events controlling synthesis of papovavirus proteins.

Another level of control is potentially available to viruses which integrate their DNA into host DNA, as transcription of viral information could be initiated from an adjacent cellular promoter. Under these circumstances early protein synthesis would be absolutely dependent upon the integration event. While this is still only a possibility, the synthesis of late mRNA of papovaviruses takes place from free DNA and hence uses a viral promoter site.

Adenoviruses

Despite the fact that adenoviruses have over 6-fold more DNA than papovaviruses, their control strategies are remarkably similar: early protein synthesis, DNA replication, late protein synthesis. However, both cellular RNA polymerases II and III are employed in transcription of both early and late mRNAs.

Study of the adenovirus system has recently been responsible for the staggering discovery that mRNAs are encoded by non-contiguous regions of the viral DNA (Fig. 8.11). Once again study of animal viruses sheds light upon processes unique to eukaryotic cells. This finding implies that the initial RNA transcript is specifically cleaved and then appropriate parts of the molecule joined with an RNA ligase. Why events should evolve to be so complex has not been explained. Cleavage has long been known in the processing of the 45S ribosomal RNA which is precursor to the mature 28 and 18S rRNAs but the ligation step is revolutionary (although RNA ligases have been found earlier in both prokaryotic and eukaryotic cells).

In the adenovirus-infected nucleus, control procedures impinging upon potential mRNAs reach great complexity. Not only is there the cleavage and ligation mentioned above, but polyadenylation, capping, methylation and

127

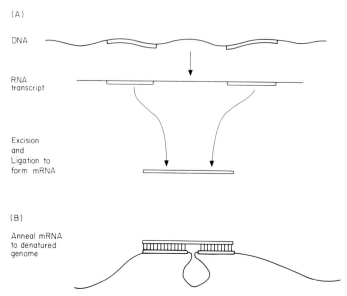

(A)

DNA

RNA
transcript

Excision
and
Ligation to
form mRNA

(B)

Anneal mRNA
to denatured
genome

Figure 8.11 (A) Production of an adenovirus or (an SV40 virus) mRNA. (B) The 'R-loop' of DNA, formed by annealing with genomic DNA, is seen by electron microscopy and provides compelling evidence for the formation of mRNA from non-contiguous regions of DNA. Only one DNA strand is illustrated.

differential transport from the nucleus occur and the permeability of the nuclear membrane to newly synthesized mRNAs changes with time after infection, so that the spectrum of viral proteins being synthesized is under continuous review.

Herpesviruses

Most is known about control of herpes simplex type 1 virus which has about 4-fold more DNA than adenoviruses. Herpesviruses depart from the scheme of gene expression outlined for the papovaviruses only in that transcription of late mRNAs is independent of DNA replication. Thus when DNA synthesis is inhibited, both early and late transcripts are detected in the nucleus. However, protein synthesis in HSV 1-infected cells follows a system of induction and repression which occurs in three phases called, α, β and γ. α-proteins are the earliest to appear and induce the synthesis of β-proteins; these in turn repress the expression of α-proteins and induce β-proteins and so on. Since, as previously mentioned, all transcripts are synthesized continuously, control operates not at the transcriptional level but on translation or processing events in the nucleus. Currently the latter view is favoured.

Poxviruses

These are the only DNA viruses to multiply in the cytoplasm. They are also the largest with a coding capacity 50-fold greater than parvo- and papova-viruses. Their site of multiplication in the cytoplasm can be likened to the establishment

128

of a second 'nucleus'. Indeed virions contain many enzymatic activities which duplicate those found in the nucleus itself. The virion has its own DNA-dependent RNA polymerase as well as enzymes which cap, methylate and polyadenylate the resulting mRNAs.

Translational control too is shown during the synthesis of viral thymidine kinase, an early protein. In the cell, early viral mRNAs are very stable and yet synthesis of thymidine kinase ceases to increase after synthesis of the late class of proteins. This argues that there is direct inhibition at the translational level of the thymidine kinase mRNA.

Retroviruses

Retroviruses combine many of the properties of both RNA and DNA viruses. The DNA provirus is synthesized in the cytoplasm (Chapter 15) and is transported to the nucleus and there integrated with host DNA. mRNAs are presumably synthesized from integrated DNA (although this has not been demonstrated) and transported to the cytoplasm. Transcriptional controls operate to synthesize three mRNAs (Fig. 8.12). The virus encodes four proteins in the order *gag, pol,*

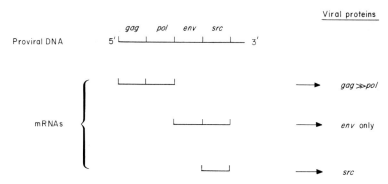

Figure 8.12 mRNAs and proteins synthesized in retrovirus-infected cells. (Not to scale.)

env and *src*. *gag* and *pol* are transcribed into a single mRNA which has a termination codon after *gag*. Thus usually only *gag* is synthesized but synthesis of *pol* (reverse transcriptase) is permitted by the appropriate suppressor tRNA and read-through. In the cell, *gag* gene products far exceed those of *pol* in amount, presumably reflecting the amounts of structural protein and enzymatic activity required for successful multiplication. The structure of mRNA coding for *env* is similar as it contains the information for *src*. However *src* is not synthesized from this structure (which thus has a similar strategy to the alphavirus virion RNA) but is made on a monocistronic message.

9 The process of infection:
IV the assembly of viruses

In an infected cell the viral-specified protein and nucleic acid are synthesized separately and consequently must be brought together and assembled into mature virus particles. There are two ways in which this may occur. Either the viruses can *self-assemble* in a manner akin to crystallization whereby the various components spontaneously combine to form virus particles, since this represents a minimum energy state; alternatively, the viral genome may specify certain morphogenetic factors which are not structurally part of the virus but whose presence is required for normal assembly. It is now clear that both forms of assembly occur, the former being prevalent among the structurally simpler viruses whereas the latter has been observed in the tailed bacteriophages which are structurally more complex.

SELF-ASSEMBLY

Absolute proof of self-assembly requires that purified viral nucleic acid and purified structural proteins, but no other proteins, be able to combine *in vitro* to yield particles which resemble the original virus in shape, size and stability as well as demonstrating infectivity. In practice, however, the critical step in demonstrating assembly *in vitro* is really disassembly. The virus must be disassembled in such a way that the released sub-units retain their ability to reassemble in a specific manner relating to their origin. Consequently, proper assembly constitutes the test for proper disassembly! In an ideal experiment the virus would be disassembled in such a manner that the protein coat separates into its constituent monomers which would not be denatured. This goal has not yet been achieved but methods of protein isolation are available that keep damage to the protein to a minimum and any denaturation reversible.

Assembly of tobacco mosaic virus: introduction

Although self-assembly is known to occur in the formation of other biological structures, e.g. flagella, the best studied example is undoubtedly the *in vitro* reconstitution of tobacco mosaic virus (TMV). This virus consists of a long helically wound filament of RNA embedded in a framework of small, identical protein molecules ('A' protein)† which are also arranged helically (Chapter 2). As already outlined in Chapter 1, TMV can be disassembled to yield protein and RNA components which can be reassembled *in vitro* to yield active virus. However the isolated protein, free of any RNA, can also be polymerized into a helical structure indicating that bonding between the sub-units is a specific

† TMV protein is usually obtained by mild alkali treatment of pure virus. Alkali destroys the RNA leaving 'alkalischer' of 'A' protein.

property of the protein. Thus the most likely model for the assembly of the virus would be for the protein molecules to arrange themselves like steps in a spiral staircase enclosing the RNA as a corkscrew-like thread. Recent research, however, has indicated that the assembly of TMV is a much more complicated process than that suggested above.

In solution, TMV protein forms several distinct kinds of aggregates depending on the environmental conditions, particularly ionic strength and pH (Fig. 9.1).

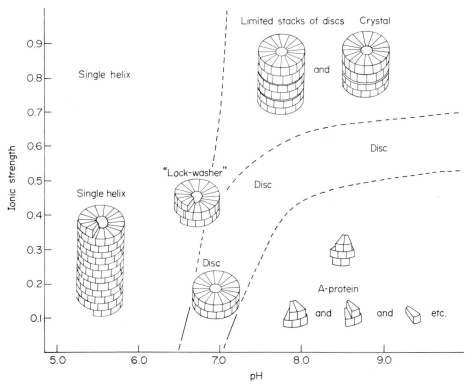

Figure 9.1 Effect of pH and ionic strength on the formation of aggregates of TMV protein.

Of these, the disc structure is considered the most important since it is the dominant aggregate found under physiological conditions. The disc structure is also found with protein from several other plant viruses such as tobacco rattle virus and barley stripe mosaic virus which are unrelated to TMV. A rod-like structure built of stacked rings could well arise as a variant of the normal helical structure; the protein sub-units would have a similar bonding pattern to that in the virus, but in the absence of RNA-protein interactions there could be small local differences in packing so that turns of the helix are transformed into closed rings. Since there are 17 sub-units per ring which is close to the 16.34 per turn of the viral helix, the lateral bonding between the sub-units is probably very similar. However, if the stacked disc structure is a true variant of the helical structure then successive rings would face in the same direction as do successive turns of the helix. That is, the structure would be polar like the helix and polar aggregation of rings would be expected to continue indefinitely and not stop at the two ring structure.

This problem was resolved by careful analysis of electron micrographs which revealed that there is a slight axial perturbation at the edges of the sub-units such that a double disc presents a marginally different aspect to the underside of a further disc adding to it (Fig. 9.4). When discs are taken to a lower pH, aggregation progressively occurs with the appearance of short rods made up of imperfectly meshed sections of two helical turns which after many hours can anneal to give the regular virus-like structure (Fig. 9.1). An explanation for this comes from titration studies which showed that TMV contains two ionizable groups per sub-unit titrating with a pK close to 7.0. Since the protein contains no histidine and no terminal amino groups, it has been suggested that these are carboxylic acid groups titrating abnormally. This would result from the assembled helix constraining two pairs of carboxylate groups per sub-unit to be close to one another. By reducing the pH the electrostatic repulsion between the carboxylate groups is reduced, thus converting the disc into a 'lock-washer'. For the purposes of reconstitution of the virus, the conversion to the helical mode could be brought about by interaction of the protein with RNA which provides the additional energy necessary for the stabilization of the helical form.

When A-protein sub-units and small aggregates are mixed with TMV RNA

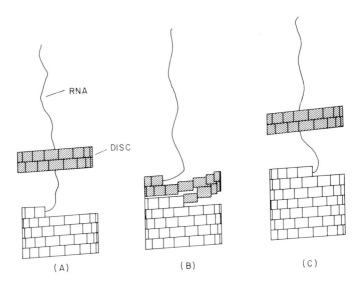

Figure 9.2 Simple model for the assembly of TMV. See text for details.

polymerization is slow and formation of virus requires about 6 hours. However, when discs are mixed with RNA under the same conditions, polymerization is rapid and mature virus forms within 5 minutes. Addition of small aggregates as well as discs to the RNA does not increase the rate of polymerization. These results strongly suggest that discs are the normal source of protein for the assembly of viruses. A possible model for the assembly of TMV is shown in Fig. 9.2 (but note that an alternative model is presented later). A disc is added to the growing helix (A) and this is converted to the lock-washer form (B) as a result of neutralization of the juxtaposed carboxyl groups by interaction with the viral RNA. Following conversion to the lock-washer form, the sub-units progressively unroll entrapping the RNA until the structure is ready to receive another disc (C).

The involvement of discs rather than individual sub-units is not too surprising since the first turn of a helical structure is hard to build thus necessitating nucleation. One solution is to have a firmly-bonded pre-existing ring which could also be transformed into the first turn of the helix. A single ring tends to be an unstable structure because there is only a single bond between adjacent sub-units. A disc made of two rings is considerably more stable because the ratio of bonds to sub-units is much greater. Furthermore, when discs are used rather than single sub-units the assembly process is less likely to be affected by short unfavourable nucleotide sequences in the RNA.

The assembly of TMV: bidirectional growth

Until recently it was thought that assembly occurred by a polar mechanism starting at the 5' end of the RNA. This conclusion had been inferred from two sorts of experiments. The first involved digestion of TMV RNA with exonucleases. Digestion with spleen phosphodiesterase, a 5' to 3' exonuclease, resulted in the release of up to 50 nucleotides and inhibited assembly of particles. Venom phosphodiesterase, a 3' to 5' exonuclease had no effect. However, it has since been shown that TMV RNA has a 5' terminal 'cap' (see page 55) and this would prevent digestion of TMV RNA by spleen phosphodiesterase which requires a free 5' hydroxyl group. The inhibition of assembly seen in these experiments was probably due to the activity of a contaminating endonuclease. The second sort of experiment which supported unidirectional assembly involved electron microscopy of partially reconstituted virions. These consistently showed protein attached to one end of an RNA chain and as reconstitution progressed the exposed RNA became shorter in length. Protein was never seen attached to both ends and not the middle, or attached to the middle but not the ends. However, it should be borne in mind that if protein initially attached close to one end only one tail of RNA might be seen.

Recent studies have shown that upon mixing purified TMV RNA with limited amounts of viral coat protein in the form of the disc aggregate, a unique region of the whole RNA becomes protected from nuclease digestion. The protected RNA consists of fragments up to 500 nucleotides long which are found in nucleoprotein particles having a protein-nucleic acid ratio similar to the mature virus. The shortest fragments define a core about 100 residues long common to all the fragments, while larger ones are covalently extended by up to 400 nucleotides in one direction and up to 30 in the other. These data are interpreted as showing that assembly is initiated at a unique internal nucleation site on the RNA, and that growth occurs bidirectionally but at greatly unequal rates. The major direction of assembly would be 3' to 5' and consistent with this view is the finding, from sequence analysis of the protected fragments, that the nucleation site is close to the 3′ end of TMV RNA. Furthermore, sequence analysis of the initiation site suggests strongly that it exhibits a hairpin configuration (Fig. 9.3).

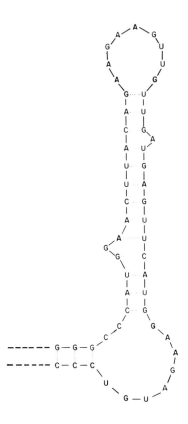

Figure 9.3 The initiation region of TMV RNA. The loop probably binds to the first protein disc to begin assembly. The fact that guanine is present in every third position in the loop and adjacent stem may be important in this respect.

Assembly of TMV: the 'travelling loop' model

This model suggests that the RNA hairpin inserts itself through the central hole of the disc into the 'jaws' between the rings of sub-units. The nucleotides in the double-stranded stem then unpair and more of the RNA would be bound within the jaws. As a consequence of this interaction the disc becomes converted to a lock-washer structure trapping the RNA (Fig. 9.4). The special configuration generated by the insertion of the RNA into the central hole of the initiating disc could subsequently be repeated during the addition of further discs on top of the growing helix; the loop could be perpetuated by drawing more of the longer tail of the RNA up through the central hole of the growing virus particle. Hence the particle could elongate by a mechanism similar to initiation, only now instead

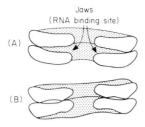

Jaws
(RNA binding site)

(A)

(B)

Figure 9.4 Diagrammatic representation of a cross-section through a disc (A) and a lock-washer (B). The arrangement of the two layers of sub-units in the disc can be likened to a pair of nut-crackers. The 'jaws' are the RNA binding site and on conversion of the disc to a lock-washer the jaws close trapping the RNA.

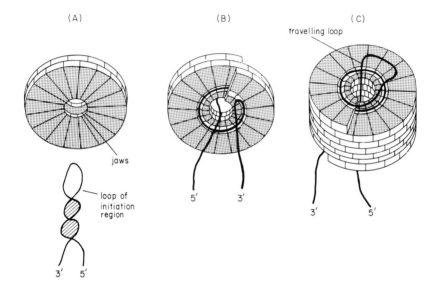

(A)   (B)   (C)

travelling loop

jaws

loop of
initiation
region

5′   3′

3′   5′

3′   5′

Figure 9.5 The 'travelling loop' model for TMV assembly. Nucleation begins with the insertion of the hairpin loop of the initiation region of TMV RNA into the central hole of the first protein disc (A). The loop intercalates between the two layers of sub-units and binds around the first turn of the disc. On conversion to the lock-washer from (B) the RNA is trapped. As a result of the mode of initiation the longer RNA tail is doubled back through the central hole of the rod (C) forming a travelling loop to which additional discs add rapidly.

of the specific initiation loop there would be a 'travelling loop' of RNA at the main growing end of the virus particle (Fig. 9.5). This loop would insert itself into the central hole of the next incoming disc, causing its conversion to the lock-washer form and continuing the growth of the virus particle.

A disturbing feature of the earlier model shown in Fig. 9.2 is that discs have to be threaded onto the RNA chain and this would obviously be the rate-limiting step. However, the 'travelling loop' model shown in Fig. 9.5 overcomes this problem as far as growth in the 5′ direction is concerned for incoming discs would add directly onto the growing protein rod. Discs would still have to be threaded on to the 3' end of the RNA and thus elongation in this direction would be much slower, as has been experimentally observed. One prediction of the 'travelling loop' model is that both the 5' and 3' tails of the RNA should protrude from one end of partially assembled TMV particles. Electron micrographs of such structures have now been observed.

The spherical plant and bacterial viruses

The most detailed study so far reported on the self-assembly and reconstitution of a spherical virus is that of cowpea chlorotic mottle virus (CCMV). This virus, which is related to broad bean mottle virus (BBMV) and brome grass mosaic virus (BGMV), has a particle weight of 4×10^6 daltons, contains about 24% RNA and has a protein shell composed of 180 sub-units. The formation, in a stoichiometric mixture of initially separated CCMV RNA and protein, of a nucleoprotein with the infectivity of CCMV affords proof of the ability of this virus to self-assemble. BGMV and BBMV have also been shown to reassemble and form hybrids. Hybrids of BBMV protein and CCMV RNA can also be prepared as can virus containing both CCMV and BBMV proteins in the one coat.

There is no evidence to suggest that specificity exists between particular nucleic acids and proteins in capsid formation with spherical plant viruses, and the formation of hybrid viruses noted above rules against such a hypothesis. Furthermore CCMV protein can be used to encapsulate RNA from other sources such as TMV, bacteriophage f2 and soybean ribosomes and those particles containing TMV-RNA are infectious. Single-stranded circular DNA from bacteriophage S-13 and double-stranded calf thymus DNA have also been encapsulated. From the foregoing it is clear that it is possible for heterologous nucleic acid to be packaged in viral protein and this could possibly provide a method for introducing foreign genetic material into plant cells.

The 180 sub-units of CCMV are normally arranged into 32 morphological units on the surface of the particle such that there are 20 hexamers and 12 pentamers at the vertices of an icosahedron of 250 Å diameter. When the conditions for reaggregation are changed, among the products formed are ellipsoids, small spheres of 160 Å diameter, double-shelled spheres of outer diameters 250 Å or 340 Å, narrow tubes 160 Å diameter and wide tubes 250 Å in diameter. Furthermore, many of these variants have less RNA than the original virus. The formation of these aberrant forms is of interest because the normally spherical viruses represented by the polyoma-papilloma group sometimes form tubes *in vivo* and they too are deficient in nucleic acid.

The first reconstitution experiments with spherical bacteriophages yielded particles which had the same size, density, RNA content and electron microscopic appearance as the original phage. However they sedimented slower than phage formed *in vivo* and could not adsorb to bacteria. These defective particles formed *in vitro* resemble those formed in restrictive hosts infected with phage carrying an amber mutation in the A-protein cistron. It was therefore argued that the A-protein is missing in both types of particles. Indeed, when the A-protein is added in stoichiometric amounts to the reconstitution mixture, infectious virus is formed.

It is not yet clear at which stage the A-protein enters the assembly process. The phage-like particles which are formed in the absence of A-protein might be precursors of infectious phage or, alternatively, the A-protein might be needed early in the assembly process for proper RNA folding and packaging. The experiment described below suggests the latter possibility.

If infected *E. coli* cells are deprived of histidine at a time when some RNA replicase has been formed, synthesis of phage RNA and coat protein continues since neither contains histidine, but A-protein synthesis ceases and defective particles are assembled. When histidine is added back to the culture, A-protein synthesis was restored as measured by the appearance of biologically active phage. Appropriate labelling experiments revealed that these were newly-synthesized phage and that A-protein could not convert the defective particles formed in its absence into viable phage.

Assembly pathways for spherical viruses

The two groups of spherical viruses just discussed exist as a complex of RNA with 180 coat protein molecules and this obviously cannot be synthesized in a single step. Intermediates in virus assembly must exist and at least three pathways can be envisaged.

1. The protein molecules first react with each other forming loosely bound protein aggregates that end up in an empty shell which in a final step is filled with RNA. However, the catalytic activity of RNA in particle formation *in vitro* makes this model unlikely.
2. Free RNA has roughly the size and shape of the RNA as it exists inside the capsid. Around this mould the protein sub-units assemble and form the capsid. Since RNA exhibits a catalytic effect regardless of its source and base composition, e.g. f2 RNA encapsulation by CCMV protein, and since these different RNA molecules are unlikely to have the same mould, this model can probably be rejected.
3. The RNA forms an initiation complex by reacting with a few protein sub-units or, in the case of the RNA phages, with the A-protein. RNA condensation occurs either during this step or in the subsequent aggregation of additional sub-units that yields the capsid. Consistent with this model is the isolation of complexes consisting of a few protein sub-units and one RNA strand.

Assembly *in vivo* is probably different in some respects from that occurring *in vitro* especially with regard to local concentrations of the components and their nascent states. Thus single-stranded RNA is used in the *in vitro* reconstitution of bacteriophages while *in vivo* the pool of single-stranded RNA is very small if it exists at all. Thus translation and self-assembly might be coupled and regulate each other. A model for such a process is shown in Fig. 9.6. The RNA replicase has replicated the single-stranded tail to the extent that the A-protein cistron and the coat protein cistron are available for translation. One molecule of A-protein

137

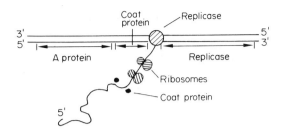

Figure 9.6 A model for the coupling of transcription, translation and protein synthesis in the regulation of RNA phage development.

and a few molecules of coat protein have already been synthesized and are bound to the RNA. The ribosomes continue to translate the RNA. The coat protein molecules situated close to the RNA replication fork and RNA replicase delay further RNA synthesis, and RNA replicase production is thus repressed at the level of RNA transcription. More coat protein is synthesized that binds to the RNA on different sites except for the coat protein cistron that is occupied by ribosomes. Then the RNA replicase resumes its activity and releases the completed RNA (+) strand. However, the completed + strand remains repressed for RNA replicase synthesis, but this time at the level of translation. Finally the ribosomes fall off and the capsid is completed.

DIRECTED ASSEMBLY

Assembly of bacteriophage T4

As we saw in Chapter 2, the structure of the tailed bacteriophages is considerably more complex than that of the viruses we have just been considering. Extensive studies on these bacteriophages, particularly T4 and λ, have revealed that non-structural proteins can direct assembly of viruses. The results obtained from the study of assembly of both phages are analogous but since the greatest body of information exists for T4 we shall confine our subsequent discussion to this virus.

The first advance in elucidating the morphogenesis of T4 came with the isolation of conditional lethal mutants (see page 13). When non-permissive E. coli cells are infected with phage carrying amber mutations in any of the structural genes, phage particles are not produced. However, electron microscopic examination of lysates of these infected cells reveals the presence of structures readily recognizable as phage components. Thus, cells infected with mutants in genes 34, 35, 36, 37 or 38 accumulate phage particles which appear normal except for the absence of tail fibres. We can thus conclude that these genes code for proteins involved in tail-fibre assembly and that attachment of tail fibres occurs late in the assembly process. Since tail fibres also accumulate in cells infected with mutants in genes 34, 35, 36 and 38, but not gene 37, this latter gene must be the structural gene for the major fibre protein. Extension of this technique to cells infected with other mutants led to the identification of the function of many T4 genes (Table 9.1) as well as suggesting that the head, tail

Table 9.1 Phenotypes of T4 mutants as determined by electron microscopy of infected cell lysates

MUTANT GROUP	GENES	PHENOTYPE	CONCLUSIONS
Y	20, 21, 22, 23, 24, 31, 40.	Fail to produce heads or generate aberrant heads	These genes involved in determination of head shape
X	2, 4, 50, 64, 65, 16, 17, 49.	Produce heads that appear normal by electron microscopy but extracts in-active when tested by *in vitro* complementation	These genes involved in head maturation subsequent to shape acquisition
	5, 6, 7, 8, 10, 11, 25, 26, 27, 28, 29, 51, 53.	Fail to produce baseplates	Required for synthesis of baseplates
	3, 15, 18, 19.	Fail to produce complete tails	Required for assembly of tail
	34, 35, 36, 37, 38, 57.	Fail to produce tail fibres or produce only half-fibres	Required for assembly of tail fibres
	13, 14.	Normal heads and tails produced but no mature particles	Required for head-tail joining

and tail fibres were all synthesized independently of each other.

The next advance came a few years later when it was discovered that the morphogenesis of T4 could be made to occur *in vitro*. In one experiment, purified fibreless phage isolated from cells infected with a tail fibre mutant (genes 34-38) were mixed with an extract of cells infected with a mutant (gene 23) which cannot synthesize heads. Infectivity rapidly increased by several orders of magnitude indicating that the gene 23 mutant extract was acting as a tail fibre donor whilst the other extract supplied the heads (Fig. 9.7). This type of experiment is analogous to a genetic complementation test but is different in that it occurs *in vitro*.

The foregoing analysis provides only limited information concerning the sequence of assembly reactions. To determine the exact sequence the precursor structures have to be isolated prior to the complementation tests. For example, free baseplates can be isolated from gene 19, gene 48 and gene 54 defective extracts by zone sedimentation and after isolation they still retain their *in vitro* complementation activity. By determining the ability of each to complement the other two unfractionated extracts it is possible to establish the sequence in which the three gene products interact with the baseplate. If isolated gene 19 defective baseplates complement a gene 54 defective extract, the gene 54 product must have carried out its function by interacting with the baseplate in the absence of the gene 19 product, i.e. the gene 54 protein acts prior to the gene 19 protein. If, moreover, in the converse experiment isolated gene 54 defective baseplates do not complement a gene 19 defective extract then the gene 19 product cannot act prior to the gene 54 product and the sequence 54, 19 is established.

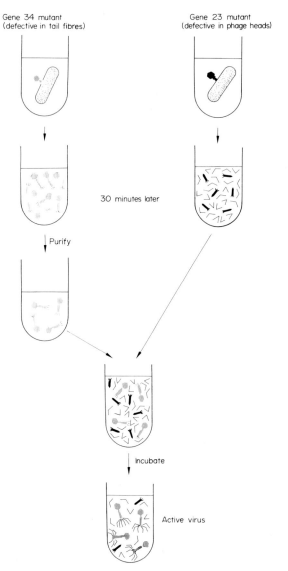

Figure 9.7 *In vitro* complementation between two mutants which are blocked at different stages in morphogenesis.

Using these techniques in conjunction with analysis of proteins in polyacrylamide gels it has been possible to determine the way in which heads, tails and tail fibres are constructed prior to their assembly into mature virions.

Head assembly

The way in which the T4 head is assembled can be deduced from the properties of mutants classified in groups X and Y (Table 9.1). A more detailed summary

Table 9.2 Phenotypes of T4 mutants belonging to group Y

Mutant Gene	Phenotype
23	No head structure
31	Produce lumps (aggregates of head protein associated with inner cell membrane)
22	Formation of multi-layered polyheads consisting of several concentric tubes approx. 20 times length of normal head
20,40	Produce single-layered polyheads
21,24	Produce tail particles (head-like structures which lack sharp edges and vertices)

of the properties of the group Y mutants is given in Table 9.2. The order in which the Y group genes act during head morphogenesis can be defined by considering the relative complexities of these aberrant structures in terms of the information needed to specify them. The fewer genes needed to specify a given structure, the simpler it must be and the earlier it must lie relative to the assembly pathway. Based upon this concept, the first stage of head morphogenesis is as shown in Fig. 9.8. Note, however, that the structures shown are *not* intermediates

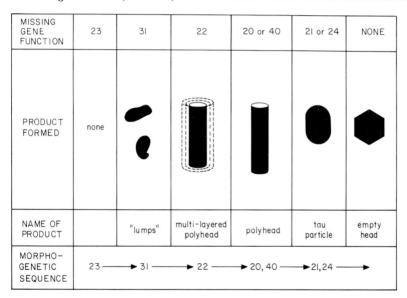

Figure 9.8 Functions of morphogenetic genes of group Y. Note the structures formed are *not* intermediates in assembly but are products of abortive assembly.

in assembly but represent the products of abortive assembly. Since multi-layered polyheads would appear to be more complex than single-layered polyheads the reader can be excused for thinking that genes 20 and 40 should act before gene 22. The relative order of their actions in the hierarchy of assembly is suggested by

141

the observation that the double mutants 22⁻20⁻, 22⁻40⁻ and 22⁻24⁻ all produce the multi-layered polyheads characteristic of infection with gene 22 mutants. This suggests that the multi-layered polyhead results from the absence of gene 22 protein which is needed for the formation of single-layered polyheads. The most likely explanation is that gene 22 protein is required for initiation of polyhead formation and in its absence it is easier to start another layer on an existing polyhead than to initiate new polyheads.

Electrophoretic analysis of the proteins of T4 heads identifies the gene 23 product as the principal structural protein although other proteins are present. However, in mature heads the product of gene 23 is present in a cleaved form, P23* which is 17% shorter than the primary gene product, P23. Cleavage of P23 is prevented by mutation in any one of the Y genes suggesting that polyheads and tau particles contain P23 rather than P23*. This has been confirmed experimentally so cleavage must occur on a precursor head structure and may be involved in transition from one head precursor to another.

Maturation of the head

When infected cells are given a pulse of ³H-leucine at 13-14 minutes after infection and the phage precursors extracted, label is first found in particles sedimenting at 400S. This label can be chased into particles sedimenting at 350S, then 550S and finally enters mature heads sedimenting at 1100S. The particles sedimenting at 400S, called proheads I, resemble tau particles and contain largely gene 23 protein. Proheads I are converted to proheads II which contain P23* instead of P23. The difference in the rates of sedimentation of prohead I and prohead II is about what would be expected from a reduction in weight of the particle when P23 is cleaved to P23*. The 550S particle, termed prohead III, is about half-full of DNA and appears to represent an intermediate between the 350S particle and the mature head. Its isolation as such implies that head filling may take place in two stages: first, half of the DNA is inserted rather rapidly and then the second half is inserted. Thus packaging does not take place continuously since this would generate precursor heads at all stages of fullness. These stages are shown diagrammatically in Fig. 9.9.

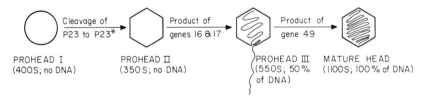

Figure 9.9 Maturation of the head of bacteriophage T4.

The dimensions of the T4 head (1000 Å × 650 Å) correspond to an internal volume of $2.5 \times 10^{-4} \ \mu m^3$ assuming the head proteins themselves occupy no space. Since the genome of T4 is 56 m long it probably occupies a minimum volume of approx $1.8 \times 10^{-4} \ \mu m^3$. Clearly, the T4 genome must be compactly packed into the T4 head. Exactly how this is achieved is not yet clear. Three genes, 16, 17 and 49, have been implicated in DNA packaging. Genes 16 and 17 are required to initiate packing, i.e. the conversion of proheads II to proheads III, and gene 49 to complete packaging. However, the mode of action of these proteins is not known.

Assembly of the tail

The T4 tail includes the head-tail connector, the core to which it is attached, the surrounding sheath, and the baseplate and tail fibres. These components control the initial stages of adsorption outlined on page 70. The pathway of tail morphogenesis is shown in Fig. 9.10.

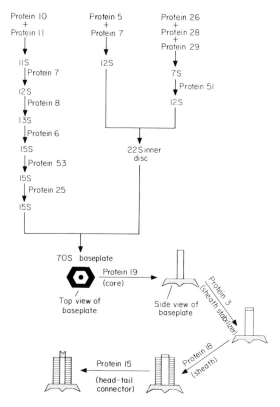

Figure 9.10   The pathway of tail assembly. The precursors of the baseplate are not individually named and only the sedimentation value is given. Note that the proteins involved in baseplate formation are not necessarily structural proteins.

Assembly of tail fibres

Mutants in any one of six genes, 34, 35, 36, 37, 38 and 57, prevent the synthesis of tail fibres. Anti-T4 serum contains antibodies that react with three fibre components called A, B and C. By determining which gene products are needed for the production of each antigen, and then identifying in extracts from defective mutants the position in a sucrose gradient at which each antigenic activity is found, the structures represented by each antigen have been identified. The way in which these are assembled into complete tail fibres is depicted in Fig. 9.11.

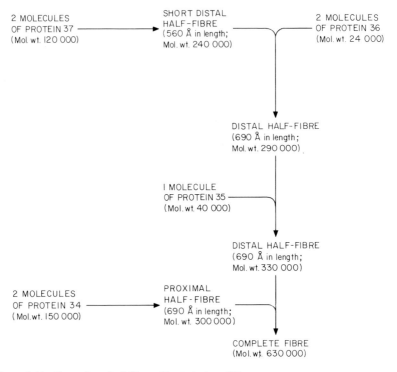

Figure 9.11 Formation of tail fibres of bacteriophage T4.

Mutants requiring co-factors for adsorption map in gene 34. Since in the absence of co-factors the fibres remain bound to the sheath, antigen A (the product of gene 34) is probably responsible for the interaction of the fibre with the particle. Host-range mutations map in gene 37 which specifies antigen C. Since only half-fibres containing antigen C can bind to T4 receptor material isolated from sensitive bacteria, antigen C must be responsible for making bacterial contact. The function of gene 57 is not clear but it may play a regulatory role since mutants in gene 57 produce fibreless particles on infection.

The overall assembly process

The way in which the different components of T4 virion are assembled into infectious particles is shown in Fig. 9.12. Although the products of genes 13 and 14 must be functional, to permit heads to join to tails, they are not involved in the joining process itself. Heads spontaneously join to tails so genes 13 and 14 must be involved in some modification of mature heads which is a necessary prerequisite to tail addition. By contrast, the tail fibres do not spontaneously join to the base-

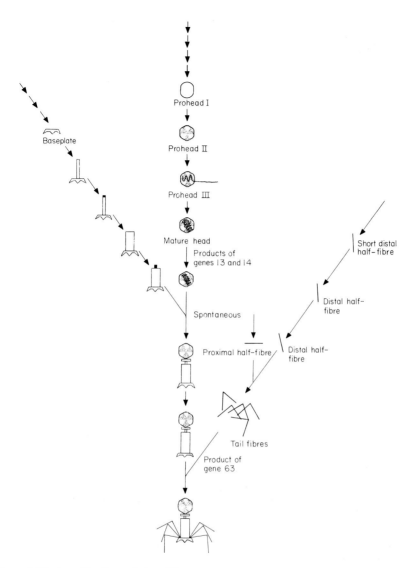

Figure 9.12 Assembly of bacteriophage T4.

plate but require the active participation of the product of gene 63. The mechanism of this catalysis is not known.

The way in which a head vertex with 5-fold symmetry stably and spontaneously unites with a tail with 6-fold symmetry (see p. 39) is not clear. Nor is it clear why tails only attach to one vertex. It is possible that one end of the DNA protrudes through a vertex, disturbing the 5-fold symmetry and this promotes tail addition. Indeed, protrusion of the DNA a short way into the tail may be a necessary structural feature for successful injection following contact of the phage with a susceptible bacterium.

Scaffolding proteins

The correct assembly of many spherical viruses, and in some cases the heads of tailed phages, appears to require the help of proteins not found in the mature virion. Such proteins are sometimes referred to as scaffolding proteins. During the morphogenesis of phage P22 about 250 molecules of scaffolding protein catalyse the assembly of 420 coat protein molecules into a double shell prohead containing both proteins. Upon encapsidation of the DNA all the scaffolding protein molecules are ejected from the prohead and these then take part in further rounds of prohead assembly. The way in which DNA replaces the scaffolding protein is not known but represents a fascinating exchange reaction. In the case of T4 the scaffolding protein, which is the product of gene 22, is removed from the prohead by proteolysis at the time of DNA encapsidation. During adenovirus assembly proteins are also lost at the time of DNA entry but it is not known if these proteins are destroyed, as with T4, or recycled as occurs during P22 assembly. There is some evidence that, like P22, herpesviruses and poxviruses recycle a scaffolding protein.

The gene B protein of ϕX174 is not found in the mature virion but functions in assembly in a catalytic role. As such it does not form a long-lived complex with structural proteins and so cannot be isolated. In ϕX174-infected cells gene F protein aggregates into pentamers as does gene G protein and in the presence of gene B protein these two structures form a new larger aggregate, possibly a vertex of the icosahedron.

The assembly of poliovirus may also be mediated by non-structural proteins. It is possible to extract from poliovirus-infected cells a protein which sediments at 13-14S and which has all the properties expected of a capsid precursor. Cytoplasmic extracts of infected cells promote the assembly of these 14S sub-units into a structure which now sediments at 73S and which is identified as RNA-free virus. The addition of similar cytoplasmic extracts from uninfected cells results in little or no assembly. These results might be interpreted as evidence for a maturation factor present only in infected cells. However, when 14S particles are sufficiently concentrated they apparently self-assemble into empty capsids in the absence of any extract from infected cells. Moreover, the kinetics of self-assembly are similar to those of extract-mediated assembly. The crucial difference between the two processes is simply that self-assembly in the absence of cytoplasmic extract depends far more on particle concentration than it does on the presence of extract. One possibility is that the extracts of infected cells contain membrane material modified by viral infection which adsorbs 14S protein and concentrates it.

Whether the assembly of poliovirus is really directed remains to be seen but since this virus resembles the spherical plant and bacterial viruses, e.g. f2 and CCMV, the phenomenon may be more widespread than we currently realise. There is certainly no doubt that directed assembly occurs with bacteriophage T4 and it remains to be seen if a similar process occurs with the complex animal viruses such as the poxviruses and the rhabdoviruses.

Protein cleavage

With most animal and bacterial viruses formation of mature virions requires the cleavage of larger precursor protein molecules after they have aggregated into proheads or pro-virions. In this context the conversion of P23 to P23* prior to DNA encapsidation in T4 heads has already been discussed (page 142). There are two possible explanations for the widespread occurrence of post-translational protein cleavage in the assembly of viruses. Firstly, in the absence of DNA or RNA the uncleaved proteins may form more stable aggregates than cleaved proteins. A relevant observation is that cleavage most frequently occurs just prior to nucleic acid encapsidation. Secondly, protein cleavage may represent a way of restricting the addition of certain minor structural components to a particular stage in morphogenesis.

It should be noted that in cells infected with picornaviruses and togaviruses two kinds of post-translational protein cleavage occur. These have been termed *formative* and *morphogenetic* and the distinction between the two is best illustrated with poliovirus. The entire genome of poliovirus is translated as a single giant polypeptide which is cleaved as translation proceeds into three smaller polypeptides (see page 121). One of these is NCVP1 (*Non Capsid Virion Protein*) which is the precursor to all four virion proteins. NCVP1 is derived from the 5′ end of the genome and is synthesized completely before being cleaved, suggesting that folding is necessary for cleavage. Secondary cleavage of NCVP1, gives rise to VPO, VP1 and VP3 which are found associated with each other in infected cells as 5S and 14S aggregates and empty capsids. The 5S complex is a precursor of the 14S complex which in turn is a precursor of empty capsids. After RNA is encapsidated VPO undergoes cleavage to yield the virion proteins VP2 and VP4. These steps are summarized in Fig. 9.13.

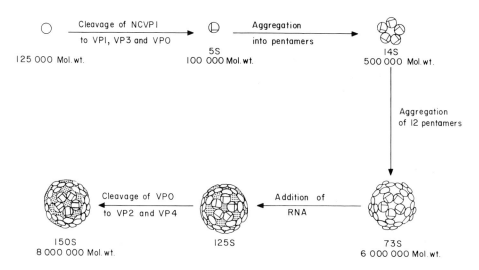

Figure 9.13 Summary of the steps involved in the assembly of poliovirus.

The initial cleavage of the primary translational product into three smaller peptides is considered to be 'formative' cleavage. This process is probably a mechanism used by animal cells as a substitute for internal initiation of protein synthesis on a polycistronic message. The cleavage of NCVP1 to VPO, VP1 and VP3 and the later cleavage of VPO to VP2 and VP4 are considered to be two examples of 'morphogenetic' cleavage.

The assembly of enveloped viruses

A large number of viruses, particularly viruses infecting animals, have a lipid envelope as an integral part of their structure. These include the herpesviruses, togaviruses, rhabdoviruses, myxoviruses, paramyxoviruses, coronaviruses, arenaviruses, oncornaviruses, poxviruses, iridoviruses and a number of bacteriophages (see Table 2.2 and Chapter 18). Herpesviruses replicate in the cell nucleus although the viral proteins are synthesized in the cytoplasm and transported back into the nucleus. After assembly of the nucleocapsid, the virus buds off from the nuclear membrane and thus becomes enveloped. Prior to the budding process the membrane is modified by incorporation of viral-specified proteins which are subsequently glycosylated. Very little is known about the assembly of the lipid-containing phages and the iridoviruses but it appears that the envelope is *not* incorporated by a budding process. In this respect they resemble poxviruses whose morphogenesis has been studied extensively by electron microscopy of infected cells. Particles initially appear in thin sections as crescent-shaped objects within specific areas of cytoplasm called 'factories' and even at this stage they appear to contain the trilaminar membrane which forms the envelope. The crescents are then completed into spherical structures, DNA is added, and then the external surface undergoes a number of modifications to yield mature virions.

The remaining enveloped viruses acquire their envelope by budding from the cytoplasmic membrane, as opposed to the herpesviruses which bud from the nuclear membrane. Four events leading to maturation have been identified. First the nucleocapsids form in the cytoplasm. Secondly, patches of cellular membrane incorporate viral glycoproteins. Thirdly, the nucleocapsids become aligned along the inner surface of the modified membrane and finally bud from the cell. During the budding process no host membrane proteins are incorporated into the viral particles although the bulk of the lipid in the envelope is derived from the host's normal complement of lipid.

SOME UNRESOLVED QUESTIONS

One question about virus assembly which was alluded to earlier, is the mechanism whereby DNA is packaged into virions. Early in infection the DNA exists as long fibrils but these subsequently condense prior to incorporation. The factors inducing DNA to condense are obscure. *In vitro,* T4 DNA can be made to condense by mixing it with a neutral polymer such as polyethylene oxide provided the salt concentration is high enough to minimize internal repulsions. Treated in this way the sedimentation coefficient of the DNA increases dramatically from 80S to 600S.

Another perplexing question is the mechanism whereby viruses with segmented

148

genomes, e.g. reoviruses and orthomyxoviruses, acquire their full nucleic acid complement during assembly. The biochemical mechanisms by which maturing virions bud from host membranes remain mysterious largely because they are not readily amenable to experimentation. Finally, the reasons why some viruses assemble in the nucleus are not understood.

10 Lysogeny

Shortly after the discovery of bacteriophages by Twort and d'Hérelle, strains of bacteria were isolated which appeared to 'carry' phage in the sense that phages were always present in culture fluids of these strains. The 'carrying' strain itself was never sensitive to the 'carried' phage but many related strains were sensitive. Although repeated single-colony isolation failed to free the carrying strain of phage, as did heating or treatment with anti-phage serum, many workers refused to believe that *lysogens,* as these carrying strains are called, represented anything other than contamination. By 1950, two more properties of lysogens had been discovered. Firstly, lysogenic bacteria are able to adsorb the phage which they 'carry', but lysis does not ensue. Neither does lysozyme lysis liberate phages from lysogens. Secondly, if the phage from a lysogen is plated on a sensitive strain, lysogens of the originally sensitive strain can be isolated.

In 1950, Lwoff ended all controversy by a simple but elegant study with a lysogenic strain of *Bacillus megaterium.* He cultivated a single cell of his lysogen in a microdrop of culture fluid and watched its division under the microscope. When the cell divided, he removed one of the daughter cells as well as some culture fluid. Observation on the remaining daughter cell was continued, and each time the cell divided, one of the progeny and some more of the culture were removed. Each of the cells which had been removed was plated on agar to deter-mine if a lysogenic population grew and the culture fluid was tested for the presence of phage. Repeated experiments of this kind showed that a lysogen could grow and divide without releasing phage. Nevertheless, a filtrate of a culture of a lysogen always contains some bacteriophages. How, then, do the phages arise in a lysogenic population? Lwoff reasoned that only a small percentage of bacteria release bacteriophages. This was confirmed when he observed spontaneous lysis of some of the cells in microdrops and that this was always correlated with the presence of free phage. It will be recalled, however, that artificial lysis of lysogens does not release phage. Thus, in lysogenic strains of bacteria there must be maintained a non-infective precursor of phage, called *prophage,* which endows the cell with the ability to give rise to infective phage without the intervention of exogenous phage particles. Lysis only ensues when some of the cells are stimulated to produce phage.

Lwoff and his two students, Siminovitch and Kjeldgaard, believed that the presence of free phage in cultures of lysogens was due to the induction of prophage by external factors. They began a long search for conditions that would increase the frequency of induction. Their search was ended when they found that small doses of U.V. irradiation induced phage production in the majority of cells of a lysogenic population. After irradiation, the lysogenic cells continue to grow and divide for some time then suddenly the turbidity of the culture drops and there is a rapid increase in the number of free phage in the culture.

Thus, a lysogenic bacterium not only passes on the ability for producing phage to all of its descendants, a few of which might spontaneously lyse and liberate infective particles, but it can also be induced to lyse at will by the application of U.V. treatment. Bacteriophages which are capable of existing as a prophage within a host cell are said to be *temperate* bacteriophages. After a lysogen is induced with U.V. light, there is initially an eclipse period during which no infectious phage can be detected inside the cell. However, phage specific DNA and protein can be detected, and shortly before lysis this is assembled into mature phage. Thus, the events following induction resemble those occurring in cells infected with virulent phages such as T4.

Although a lysogenic strain of *Bacillus megaterium* was used by Lwoff, most of the subsequent studies have utilized another temperate phage, λ. In 1951, Esther Lederberg accidentally discovered that *E. coli* K12, the strain in which Tatum and J. Lederberg had discovered bacterial conjugation, was also lysogenic. Non-lysogenic derivatives had unknowingly been prepared and when these were mixed with the original K12 strain, plaques were produced, and the phage isolated from them was called *lambda*, or λ for short. A wealth of information on this phage has been produced in the last two decades, so much in fact, that an entire book has been written about it! Consequently, we shall only cover a few of the salient features here.

The integration of λ DNA

Since lysogenic and non-lysogenic derivatives of the sexually fertile *E. coli* K12 were available, it became possible to determine how the prophage behaved in genetic crosses. Early studies showed that lysogeny is closely linked to galactose fermentation. Following the isolation of Hfr bacteria and the discovery that such bacteria transfer their genetic material in an orderly fashion, it became possible to map genetic loci by measuring their times of entry into the recipient and in this way λ prophage was located between the galactose and tryptophan genes. In genetic crosses, the character of lysogeny behaves just like any other marker and it was assumed that during lysogeny the prophage DNA is in some way integrated into the host chromosome.

How, in the course of eliciting the lysogenic response, does the phage genome manage to become integrated into the bacterial chromosome? The first clue came with the isolation of λ mutants with a deletion in a particular region of the chromosome called the b2 region. These mutants were incapable of lysogenizing their host although they were perfectly capable of undergoing lytic growth. Thus, the b2 sector is involved in the genetic exchange between the λ genome and the region of the *E. coli* chromosome between the galactose and tryptophan genes, which is referred to as the lambda attachment (*att* λ) locus.

In 1962, Campbell, with remarkable foresight, suggested that crossing-over between phage and host chromosome results in insertion of the entire phage chromosome into the bacterial chromosome. He further suggested that the phage chromosome is circular and that at the moment of lysogenization the b2 region of the phage and the *att* λ region of the bacteria synapse. Thus, the phage is inserted as a linear structure into the continuity of the host chromosome by reciprocal recombination (Fig. 10.1).

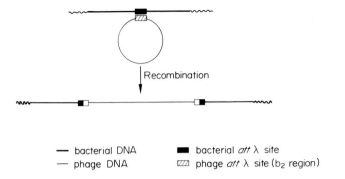

Figure 10.1 Campbell's model for the insertion of phage λ DNA into the bacterial chromosome.

Figure 10.2 Early genetic map of bacteriophage λ .

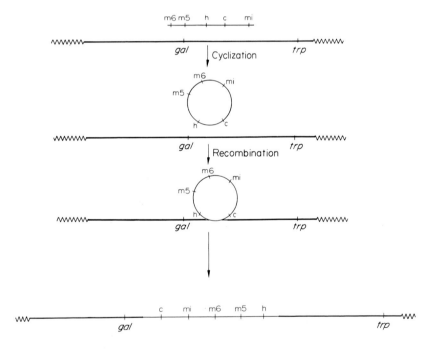

Figure 10.3 Reordering of the genes of bacteriophage λ following insertion of the phage DNA into the bacterial chromosome. The symbols *gal* and *trp* represent the galactose and tryptophan operons respectively.

Genetic studies on λ had yielded a linear map in which the b2 region and genes c and h were midway between genes mi and m6 which were located at either end (Fig. 10.2). If Campbell's hypothesis was correct, then the chromosome must assume a circular configuration prior to lysogenization followed by a reordering of the genes after insertion (Fig. 10.3). When crosses were made between Hfr and F⁻ bacteria lysogenic for different genetically marked strains, a map was obtained in which c and h were at opposite ends and mi and m6 were closely linked. Thus, the predictions of the Campbell model were borne out. About this time, Hershey and his colleagues had shown that the DNA of λ has 'sticky ends' (cf. page 47) thus providing a mechanism whereby circular molecules might be formed in the infected cell. Furthermore, when *E. coli* is infected with labelled phage and the DNA extracted, some of the label can be isolated as covalently closed circular molecules.

Structure of the attachment sites

Fifteen years after Campbell proposed his model for λ integration and excision, the bacterial and phage attachment sites had been sequenced and shown to contain identical tracts of 15 base-pairs (Fig. 10.4). The actual site of the cross-over for

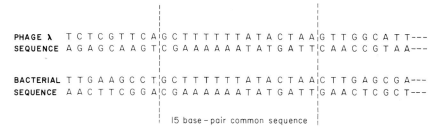

Figure 10.4   The common sequence in the bacterial and λ attachment sites.

both integration and excision must take place within, or at the boundaries of this common sequence but the precise roles of the *int* and *xis* proteins (see below) are not yet clear. However, the fact that the common sequence is rich at AT base-pairs is particularly significant in view of recent findings that negatively supercoiled DNA is required as a substrate for integrative recombination. Since negatively supercoiled DNA will tend to partially denature in regions of high A + T, there is a strong suggestion of a preference for single-stranded DNA by one or more of the components in the recombination pathway.

The excision of λ DNA

Just as recombination between the phage and host chromosome results in lysogeny, reversal of the process results in the regeneration of the circular phage chromosome and terminates in lysis. Genetic studies have revealed the presence of a battery of phage genes which specifically control integration and excision of the prophage DNA. For example, the *int* gene (integration protein) is required for the cross-over between the host and phage chromosomes. Similarly, the *int* and *xis* (excision protein) gene products are required for release of the prophage from the bacterial chromosome.

Occasional errors can occur in the process of excision, such that the region excised includes a small amount of bacterial DNA, with a corresponding piece of phage DNA deleted. If the portion of bacterial DNA is located to the left of the prophage, it may include some or all of the genes of the galactose operon, with a compensatory loss of some of the genes at the right end of the prophage map. This incomplete phage chromosome then serves as the template for replication, such that essentially all of the phage progeny issuing from that cell bear the genes for galactose utilization and have lost some of the phage genes. Reinfection of a Gal⁻ cell with such a phage, under conditions allowing lysogenization, can confer upon that cell the ability to ferment galactose, since the necessary genes become inserted into the bacterial chromosome. Such a transducing phage is defective in its own replication, since it lacks some essential phage genes and can only be propagated if a normal ('helper') phage is present to supply the missing gene functions.

By the time Lwoff had confirmed the existence of a prophage in lysogens it had already been established that super-infecting phage DNA does not replicate in a lysogen. Thus the idea arose that in a lysogen an 'immunity substance' is synthesized by the prophage which not only prevents the expression of most of the prophage genes but which is capable of diffusing throughout the cell and inhibiting the expression of super-infecting phage genomes of the same origin. Among the earliest mutants of λ which were isolated was a series which gave rise to clear plaques and these mutants could be divided into three classes (cI, cII and cIII). cI mutants behaved as virulent mutants and were incapable of lysogenizing a host except in the presence of phages carrying a functional cI gene product. However, cI mutants failed to grow in lysogens which obviously had a functional cI gene. Thus, the cI gene is most likely responsible for the production of the immunity substance.

Immunity to super-infection

Immunity is a specific property. A lysogen is immune to the phage of the same type that it carries, but is usually not immune to another, independently isolated temperate phage. Should two phages be immune to one another they are termed *homoimmune* and if not are called heteroimmune. A number of temperate phages which are related to λ but heteroimmune have been isolated and these have permitted a detailed analysis of the genetic factors involved in the specificity of immunity. In genetic crosses between two heteroimmune phages the genetic

Figure 10.5 Simplified genetic map of prophages λ, 434 and 21 (not to scale). The arrow indicates the ends of the chromosome in the mature phage particle and the dotted lines represent bacterial DNA. The area of non-homology is shown by the solid bar.

154

determinants responsible for the specificity of immunity are found to be closely linked to the cI gene. In fact, for each pair of phages tested there is a region of non-homology comprising gene cI and segments of various lengths on either side (Fig. 10.5). Furthermore, in crosses between heteroimmune phages it is possible to construct hybrids in which the immunity region is derived from one parent and the remainder of the chromosome from the other parent. Thus, from a cross between phages λ and 434, a hybrid (434 *hy*) was isolated whose chromosome is almost entirely derived from λ but which has the cI region, and thus the immunity, of 434.

The production of the immunity substance by the cI gene is analogous to the production of a repressor protein for β-galactosidase synthesis by the *i* gene of *E. coli*. Carrying the analogy further it should be possible to find mutants which correspond to the o^c mutants of the *E. coli lac* operon. Such mutants were isolated very soon after the discovery of λ, and as expected, they proved to be virulent. These λ derivatives grow on λ lysogens despite the synthesis of the cI gene product and have mutations which map in the vicinity of cI thus defining two operators, O_L and O_R, to the left and right of cI respectively. That immunity acts at the level of transcription was made clear by the observation that only λ specific mRNA synthesized in a lysogen hybridized with the cI region. This fact was confirmed by the isolation of the λ immunity substance (the 'λ repressor') by Ptashne. For this purpose, bacteria lysogenic for λ were treated with U.V. radiation to greatly reduce their endogenous protein synthesis and then super-infected with a vast excess of phage in the presence of radioactive amino acids. Under these conditions most of the residual protein synthesis is due to the formation of the cI product by the super-infecting phages since host protein synthesis had been inhibited by the pre-infection treatment and synthesis of most vegetative phage proteins was held in abeyance by the endogenous immunity repressor. In this way it was possible to isolate a radioactively labelled protein from these infected cells which was absent from cells super-infected with λ cI mutants. This protein was also capable of binding to λ DNA but not to the DNA from 434 *hy* which lacks the right and left operators of λ.

At this stage we can summarize by saying that the cI gene produces an immunity substance or repressor which acts on two operator sites, O_L and O_R, and so preventing the transcription of the majority of the λ genes (Fig. 10.6). Thus the cI gene product exerts negative control over λ development.

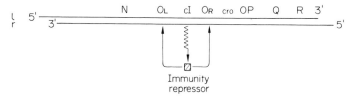

Figure 10.6 Binding of the cI gene product, the immunity repressor, to the right and left operators.

The structure of O_R and O_L

The nucleotide sequences of the repressor-binding sites, O_R and O_L, have now been determined. In order to do this advantage was taken of the fact that bound

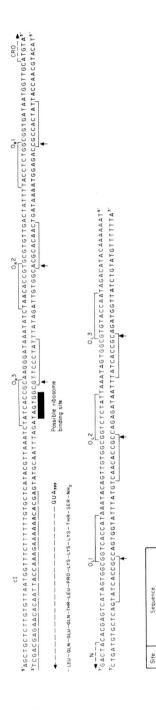

156

Figure 10.7 The nucleic acid sequence of O_L and O_R. The repressor binding sites O_L 1, 2 and 3 (left operator) and O_R 1, 2 and 3 (right operator) are set off in brackets. The vertical arrows indicate the axes of partial 'twofold symmetry possessed by each repressor binding site. The heavy line under O_R 3 indicates the six bases which are presumed to code for a strong ribosome binding site (see page 160). The start points of transcription of genes N, *cro* and *cI* are shown, as are the amino-terminal residues of the cI protein.
The table at the bottom of the figure gives a comparison of the base sequence of each of the six repressor binding sites.

Site	Sequence
O_R1	TATCACCGCCAGAGGTA ATAGTGGCGGTCTCCAT
O_R2	TAACACCGTGCGTGTTG ATTGTGGCACGCACAAC
O_R3	TATCACCGCAAGGGATA ATAGTGGCGTTCCCTAT
O_L1	TATCACCGCCAGTGGTA ATAGTGGCGGTCACCAT
O_L2	CAACACCGCCAGAGATA GTTGTGGCGGTCTCTAT
O_L3	TATCACCGCAGATGTT ATAGTGGCGTCTACCAA

repressor protects the operators from degradation by added nucleases. Both O_L and O_R turn out not to consist of a single repressor-binding sequence but of a series of three such sequences (Fig. 10.7).

The function of genes N and Q

Many prophage genes operate following super-infection of lysogens with sufficiently closely related heteroimmune phages. This is made apparent by the fact that super-infecting phages defective for these genes are complemented by the prophage. These prophage genes can thus be switched on in spite of fully persisting immunity. Thus it is not immunity *per se* which prevents expression of these genes in lysogens; rather, immunity prevents the prophage from producing diffusible molecules necessary to switch on genes. Thus lytic development also involves positive controls.

Which λ genes exert positive control? Mutations in such genes would be expected to display a recessive, pleiotropic phenotype. Mutants displaying such a phenotype have been mapped in genes N and Q. Mutants of gene Q replicate normally but they express all the late functions at a low rate. Since Q mutants produce very little messenger RNA hybridizable with the right arm of the phage chromosome, the Q gene product must act at the level of transcription. N⁻ mutants are exceedingly pleiotropic and most λ functions are not expressed in the absence of the N gene product. The N gene product also acts at the level of transcription since very little of the mRNA produced following induction of N⁻ lysogens is hybridizable with λ DNA. Since in cells lysogenized with a λ strain carrying a *ts* gene N, there is a thermosensitive synthesis of many other gene products, the pleiotropic character of N mutants must be due to the absence of the N gene product rather than to a polar effect.

However, the N and Q genes do not act independently of each other. Rather, expression of Q is dependent upon the expression of N as shown below. The late gene R codes for an endolysin whose activity can be assayed *in vitro*. When susceptible cells are mixedly infected with λN⁻ and 21*hy*Q⁻R⁻ no endolysin is produced. Thus, the N gene product of 21*hy* cannot be used by the λN⁻ in order to produce endolysin. When 21*hy*Q⁺R⁻ is used as a helper phage, the λN⁻ produces a normal amount of endolysin. Thus, the need for N in the major expression of the late genes is only an indirect one. The N gene product is only needed to the extent that it is involved in the expression of gene Q. Since the Q gene product of 21*hy* can complement that of λN⁻ it must specify a diffusible substance, probably a protein. Our increasing knowledge of λ regulation is summarized in Fig. 10.8.

The region to the left of N comprises several genes including β, *exo* (general recombination) *xis*, (prophage excision) and *int* (prophage integration and excision). These genes are positively controlled by N for, in its absence, they are not expressed. These genes are also controlled by immunity at the O_L operator. How are these two controls connected?

If a λN⁻ lysogen is infected with 434*hy* *exo*⁻, (which supplies the N product), exonuclease is *not* produced. Such a result excludes the simple idea that the N product operates merely by initiating a wave of transcription from an independent

157

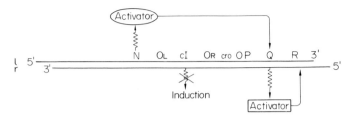

Figure 10.8 Dependence of gene Q expression on gene N expression following induction.

promoter to the left of N. However, if the λN^- lysogen is first induced, exonuclease *is* produced. Thus, the gene N product is not only required for expression of the genes to the left of it but immunity must be destroyed as well.

The most likely explanation is that the N product permits a wave of transcription, initiated within the immunity region, to progress beyond a site where it would otherwise have stopped. This 'terminator' site could either be to the right or left of N. If it were to the right of N, N would be positively regulated by itself. Since N specific mRNA is produced in cells infected with λN^-, N cannot control itself and so the terminator site must be to the left of N. This information is summarized in Fig. 10.9.

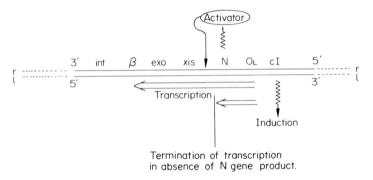

Figure 10.9 Transcription of the genes involved in excision as a consequence of gene N expression.

Expression of the genes between cI and Q (the X operon)

The region to the right of cI comprises genes with very different functions: *cro*, which has a regulatory role (see later), cII which takes part in immunity and the replication genes O and P. Unlike other genes in the X operon the *cro* gene is transcribed independently of N. However, evidence for direct positive control of these other genes by N is not as convincing as the evidence for N control of gene Q. Determining the role of N is complicated by the fact that genes cII, O and P can be transcribed in the absence of N. However, the three gene products are synthesized in much greater amounts in the presence of N than in its absence. Presumably there is a weak termination signal between *cro* and cII. Since the gene to the right of P is Q, whose transcription is dependent on the presence of N, we must postulate the presence of a second terminator between P and Q. This aspect of λ control is summarized in Fig. 10.10.

158

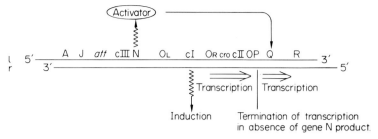

Figure 10.10 Dependence of gene Q transcription on presence of gene N product.

Summary of events during the lytic cycle of λ

1. The λ DNA is injected into the cell and transcription begins resulting in the formation of the N and *cro* gene products as well as a little of the cII, O and P gene products.

2. N exerts positive control because it is an antiterminator so there is transcription of (a) genes cIII to *att* and (b) cII to Q.

3. Q exerts positive control on transcription of S, R and A to J (remember that λ DNA circularizes following infection so genes R and A are linked).

4. The *cro* gene product exerts negative control by counteracting the effect of N. This can be shown experimentally by following the synthesis of an early enzyme under N control, e.g. exonuclease (Fig. 10.11).

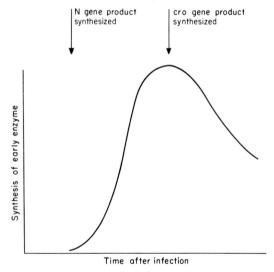

Figure 10.11 Effect of *cro* gene product on synthesis of exonuclease.

5. Transcription of O and P results in DNA synthesis which gives a gene dosage effect. As a result of the effect of *cro*, almost all the gene products synthesized will be specified by late genes, e.g. the lysis (S and R) and structural genes (A to J).

6. Phage assembly and lysis will occur.

159

The events leading to lysogeny

Earlier (p. 154) we alluded to the fact that clear plaque mutants of λ can belong to any one of three genes cI, cII and cIII. Since the λ repressor is produced by the cI gene, what is the function of the cII and cIII genes? It appears that the cI gene is transcribed in two modes. In the lysogenic state transcription begins at the promoter called P_{RM} (Fig. 10.12). By contrast, upon infection of a non-

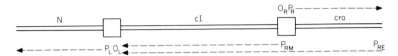

Figure 10.12 Position of the promoters for repressor establishment (P_{RE}) and repressor maintenance (P_{RM}) relative to O_R and the cI gene.

lysogenic cell, cI transcription begins about 1000 bases to the right of cI at a promoter called P_{RE} (Fig. 10.12). P_{RE} directs the synthesis of 5- to 10-fold more repressor, per genome, than does P_{RM} and provides the large burst of repressor necessary to *establish* lysogeny. Once established lysogeny can be *maintained* by the lower level of repressor synthesis directed by P_{RM}. When λ infects a cell, transcription of the N gene results in the formation, among other things, of some of the cII and cIII gene products. Both these proteins are positive regulators necessary for transcription from P_{RE} and hence their absence leads to the clear-plaque phenotype. Once lysogeny is established the cI protein will prevent transcription from Op and O_R and synthesis of the cII and cIII gene products will cease as will transcription from P_{RE}.

Why should transcription of cI initiated at P_{RE} produce more repressor than transcription from P_{RM}? Fig. 10.7 shows that the codon corresponding to the amino terminus of repressor is found immediately adjacent to the 5' terminal AUG of the cI gene. This is surprising for most mRNAs contain a leader sequence prior to the AUG or GUG translation start signals. These leaders have been found to contain short sequences that are complementary to sequences at the 3' end of 16S ribosomal RNA which promote binding of messages to ribosomes. Since the cI message transcribed from P_{RM} bears no leader sequence it may be translated at low efficiency. However, 12 bases to the right of the translational start point there is a 6 base sequence complementary to a sequence at the 3'end of 16S ribosomal RNA (see Fig. 10.7). This sequence would be present in the cI message and could function as a strong ribosome binding site.

Comparison of λ with other temperate phages

Too few temperate phages have been studied sufficiently to make detailed comparisons. It is possible to say, however, that the mechanisms of prophage maintenance and immunity are diverse among different phages as exemplified by coliphages P2 and λ and the *Salmonella* phage P22. Phage P2 seem to have only one gene which controls lysogenization and the few recessive clear-plaque mutants that have been studied all belong to one complementation group. In addition,

P2 immunity seems to involve only one operator locus and virulent mutants thus occur at a perceptible frequency. The only P2 gene, other than the repressor gene itself, which is not repressed in lysogens is the *int* gene but it is probably made quiescent when integrational recombination severs its transcription unit.

The immunity mechanism of P22 deviates from that of λ in another way. Maintenance of the P22 prophage requires the presence of two repressors produced by two different genes *mut* and *c2*. In addition, two other *c2*-linked genes, *c1* and *c3*, analogous to genes cII and cIII of λ, are required for establishment of lysogeny.

For P2, P22 and λ, stable lysogeny requires prophage insertion and in each case one gene for insertion has been found. Whereas P22 and λ integrate at a single site on the host chromosome, P2 can insert into at least nine different sites in the chromosome of *E. coli* C. The extreme example of low insertion specificity is offered by coliphage Mu 1 which can apparently insert at any point in the chromosome. The reason why Mu has the exceptional ability is not known. It does, however, have one obvious side-effect—since insertion frequently occurs in the middle of a gene it leads to gene inactivation.

THE BENEFITS OF LYSOGENY

What is the basis of the natural selection favouring lysogeny? The fact that temperate phages carry so many genetic determinants for lysogeny indicates that lysogeny does confer some advantages on the phages. One advantage would be to provide a mode of persistence which does not deplete the supply of hosts for such an obligate parasite. Another would be the opportunity for extended growth under non-selective conditions in which multiple genetic variations can occur.

Lysogeny also carries a strong selective value for the host cell since temperate phages frequently confer new characteristics on the host cell. This phenomenon manifests itself in many ways and is referred to as *lysogenic conversion*. For λ the only known lysogenic conversion is the capacity of λ lysogens to block the multiplication of a particular class of mutant (the rII mutant) of bacteriophage T4. This block involves the product of the *rex* gene and perhaps also the repressor. Since rII mutants may not be very common in the natural environment of λ, the *rex* gene probably has a more significant role to play. The *rex* gene probably influences the reproductive fitness of λ lysogens since rex^+ λ lysogens reproduce more rapidly than non-lysogens during aerobic growth in carbon-limited chemostats. Lack of the *rex* function causes λ lysogens to lose their reproductive advantage. Two facts should be borne in mind. Firstly, the human gut which is the normal habitat of *E. coli* is anaerobic and under anaerobic conditions λ lysogens reproduce more slowly than non-lysogens! Secondly, P1, P2 and Mu 1 lysogens also reproduce more rapidly than non-lysogens but do not appear to possess a gene analogous to *rex*.

Lysogenic conversion often involves exclusion of super-infecting phages. Exclusion acts *after* phage adsorption and in severe form involves breakdown of the super-infecting phage DNA. Lysogens of P2 and of P22 exclude many other phages: P22 exclusion even acts against P22. Still another way by which lysogenic conversion can act to protect the lysogen from further phage infection is shown by P1. P1 prophage confers a new restriction and modification phenotype on the host. Thus phages grown on hosts lacking P1 cannot successfully infect a P1 lysogen.

The *Salmonella* phage ε15 supplements immunity in yet another way. *Salmonella* have in their cell wall a lipopolysaccharide which constitutes the somatic O-antigen. The antigen specificity resides in the carbohydrate moiety which consists of a core with attached side chains (Fig. 10.13). The side chain structure is characteristic

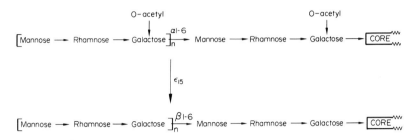

Figure 10.13 Changes in the lipopolysaccharide of *Salmonella* following lysogenization with bacteriophage ε15.

for each *Salmonella* serotype and a strain is assigned an antigenic formula on the basis of serological cross-reactions. Some of the antigenic determinants of group ε *Samonella* are controlled by phages such as ε15. For example, *S. anatum* has the antigenic formula 3,10 but after infection with phage ε15 becomes converted to *S. newington* which has the formula 3,15. If *S. newington* is cultivated in the presence of antiserum against antigen 15, reversions to *S. anatum* can be isolated and these are no longer lysogenic.

The converting activity of phage ε15 brings about three biochemical changes in the structure of the O-antigen of *S. anatum* (Fig. 10.13). These are: (1) the production of an enzyme that builds polysaccharide chains with a β-linkage between the sub-units; (2) repression or inhibition of a bacterial enzyme that generates chains with an α-linkage between the sub-units; (3) repression of a cellular enzyme that acetylates the galactose residues.

These converting functions are not essential for the phages since mutants that have lost the converting ability grow perfectly well. Yet the fact that the lipopolysaccharide of *Salmonella* is the adsorption site for bacteriophages (Fig. 4.1) suggests a possible evolutionary role of the converting genes of the phage in determining the range of bacteria which specific phages can use as hosts.

Perhaps the most interesting example of lysogenic conversion is that observed in *Corynebacterium diphtheriae*. In 1951 Freeman found that if non-toxigenic strains of Corynebacterium were infected with a phage, now called β, from virulent, toxigenic bacilli of the same species, a proportion of the survivors acquired the hereditary ability to synthesize toxin. The conversion of the non-toxigenic strain to toxigenicity was not due to the selection by the phage of a minority of toxigenic survivors but to the establishment of lysogeny. Not only was there good correlation between the frequency of lysogenization and the frequency of toxigenicity, but loss of the prophage resulted in loss of toxin production.

It is possible that a phage gene codes for the toxin itself. Alternatively, the product of a phage gene could control the expression of a host gene which codes for the toxin. A mutant phage, β_{45} has now been isolated and cells lysogenized with this phage release a non-toxic protein serologically related to toxin. This non-toxic protein has a chain length shorter than the toxin itself but cross-reacts with horse and rabbit anti-toxins. These results establish beyond any doubt that the structural gene for toxin is carried by the phage itself.

11 Interactions between viruses and eukaryotic cells

So far we have dealt with viruses as if they were able only to produce a lytic infection or to undergo lysogeny. In both cases we discussed the interactions of bacteriophages and their unicellular hosts. In this chapter we shall be concerned with the various types of interactions which occur between viruses of eukaryotes and cells in culture. These are classified here into lytic, persistent, latent, transforming and abortive infections. They have all been studied in the laboratory to determine the molecular events involved and to pave the way for our eventual understanding of the process of infection of the whole organism (Chapter 13). It should be borne in mind that a prerequisite of any of these types of infection is the initial interaction between a virus and its receptor on the surface of the host cell and any cell lacking the receptor is automatically resistant to infection.

LYTIC INFECTIONS

These are those commonly studied in the laboratory because cell killing is the easiest effect to observe and production of infectious progeny can usually be monitored without difficulty. The one-step growth curve (Chapter 1) describes the essential features of any eukaryotic or prokaryotic virus-host interaction which results in lysis. In general, host cells die and disintegrate but only in a few bacteriophage infections is this mediated through lysozyme. Viruses may kill cells by inhibiting their DNA, RNA or protein synthesis but frequently death occurs earlier than can be accounted for by any of these events. A general release of lysosomal enzymes or alteration in the ion regulation of the infected cell may be responsible in such cases.

PERSISTENT INFECTIONS

These are those which result in the continuous production of infectious virus which means that, unlike lytic infections, the host cell survives. Persistent infections result from a balance struck between the virus and its host either (1) through the interaction of virus and cells alone, (2) with the help of antibody or interferon (in animal cells only), (3) by the production of defective-interfering virus and (4) by a combination of these events.

(1) Virus + cells alone

An outstanding example of persistence of this type is seen in the infection of monkey kidney cells with a simian virus (SV5, Paramyxoviridae, Class V). The virus multiplies with a classical one-step growth curve (Fig. 11.1A) but the cells

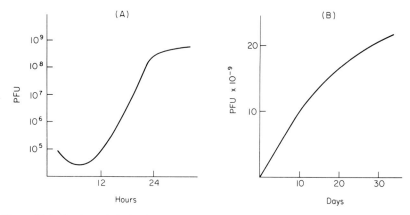

Figure 11.1 Persistent infection by SV5 of monkey kidney cells in culture. (A) Initial one-step growth curve and (B) the cumulative yield of virus from (A) over 30 days.

do not die. In fact they continue to produce virus for over 30 days (Fig. 11.1B). Infection by SV5 does not damage the cell in the sense that it does not perturb cellular DNA, RNA or protein synthesis in monkey cells. In fact virus infection makes little demand on the host's resources; for example, the amount of viral RNA synthesis is $<1\%$ of cellular synthesis (even though each cell is producing about 150 000 particles/day!). An important point to note is that SV5 lytically infects baby hamster kidney (BHK) cells. Thus the outcome of infection depends on the properties of both the virus and the host cell.

(2) Virus + cells + antibody, or virus + cells + interferon

This is a situation in which the virus would normally be lytic. However the addition of either small amounts of specific neutralizing antibody or of interferon depresses the yield of progeny virus or the extent of multiplication so that the cell population continues to survive even if a small proportion of cells die. It is suggested that this situation mimics certain sorts of persistent infections in the whole animal (Chapter 13).

(3) Virus + cells + defective interfering (DI) virus

DI viruses are produced probably by all viruses as a result of mistakes in replication which delete part of the viral genome. Propagation of these deletion mutants is favoured by a high m.o.i. but usually several successive passages are required for DI virus to reach significant proportions (see Chapter 7). Since DI viruses depend absolutely upon infectious virus for at least some essential components, they deprive a normal, infectious virus of sufficient components for its own multiplication.

An infection can be carefully engineered with critical proportions of infectious virus, cells and DI virus so that a persistent infection is achieved. In the initial stages, increase of DI virus is seen to follow that of infectious virus upon which it is dependent (Fig. 11.2). As DI virus increases there is a progressive inter-

165

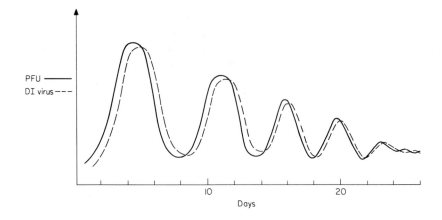

Figure 11.2 A persistent infection established with defective-interfering virus.

ference with multiplication of normal virus; its numbers decrease which in turn results in a concomitant decrease in the dependent DI virus. This cycle of events continues until the levels of virus produced fall and a persistent infection results in which there are low levels of infectious and DI virus.

LATENT INFECTIONS

Latent means existing but not exhibited. In a latent infection the viral genome and possibly various virus-coded products are present but infectious virus is not formed. Lysogeny by temperate phages (Chapter 10) is clearly a latent infection and in animal cells for example, some tumour viruses have the property of remaining latent. It is the infection rather than the virus which is latent since the properties of both cell and virus are important in establishing latency (or indeed any other type of interaction) and a virus which has a latent infection in one cell type may be lytic in another. Both bacterial and animal viruses which achieve latency do so by integrating a DNA copy of their genome into the host's DNA. This ensures that the viral genome will be replicated together with host chromosomal DNA and transmitted to daughter cells and will be protected from degradation by nucleases. It is apparent that fine molecular controls are operating to maintain the latent state since tumour virus infections may be made productive by fusing a latently infected cell with a cell susceptible to lytic infection (see next section) or more simply by seeding together such cells and so co-cultivating them. The latter circumstance probably implies that a low level of infectious virus is being produced and hence that there is a persistent rather than a latent infection, although it is difficult to exclude the possibility that co-cultivation has activated a genuine latent infection.

The presence of latent virus can sometimes be shown by immunofluorescence using antibody specific for viral proteins or by molecular hybridization with nucleic acid probes containing sequences complementary to the viral genome.

166

Cell fusion

This is a technique which arose from the observation that during the course of infection various paramyxoviruses promote the fusion of cells in culture to form multinucleate cells called *syncytia*. Later it was realized that multiplication was not essential and that large quantities of inactivated paramyxoviruses (e.g. Sendai virus) can fuse cells together (Fig. 11.3). These viruses have a protein (the F

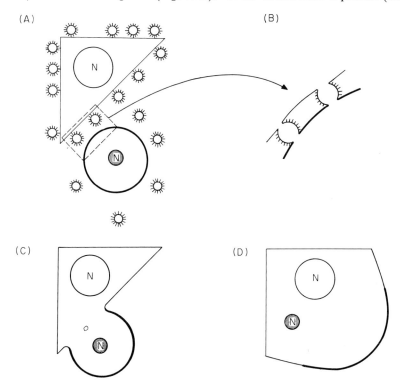

Figure 11.3 (A) to (D) are stages in the fusion of two cells through the action of non-infectious Sendai virus. In (A) plasma membranes are bridged by attached virus particles and in the next stage, at higher magnification in (B), plasma membranes of the two cells become continuous by fusion with virus. Eventually a single binucleate cell is formed. N = nucleus.

protein) in their envelope which promotes fusion but molecular details of its action are not known. Although cell fusion has been used to identify latent infections it has found wider applications in cell biology for the formation of hybrids of cells from different species.

TRANSFORMING INFECTIONS

As a result of infection with a variety of DNA viruses or with RNA tumour viruses a cell may undergo more rapid multiplication than its fellows concomitant with a change in a wide variety of other properties, i.e. it is transformed. This is accompanied by integration of the viral and host genomes as mentioned above. Chapter 15 deals in detail with transformation and other aspects of tumour viruses.

ABORTIVE INFECTIONS

These result in a reduction in the total yield of virus particles or in the particle: infectivity ratio. Both of these reflect an incompatibility between the virus and a particular host cell. A defect in the production or processing of any components necessary for multiplication, be it DNA, RNA or a protein, can give rise to an abortive infection. Defective-interfering viruses referred to above are a specific example. Another is an avian influenza virus growing in a line of mouse L cells where both the amount of progeny and its specific infectivity are reduced, probably due to the synthesis of insufficient virion RNA. Infection of other cells with human influenza viruses gives rise to normal yields of non-infectious progeny. This results from a failure to proteolytically cleave the haemagglutinin protein and can be completely reversed by adding small amounts of trypsin to the culture or to released virus. Abortive infections present difficulties to virologists trying to propagate viruses but have been used to advantage when research into the nature of the defect has furthered understanding of productive infections. In natural infections abortiveness may contribute to tissue specificity by permitting invasion only of those cells in which a productive infection can take place.

12 The immune system and interferon

Infections of multicellular plants and animals are complicated by the variety of cell types present in an individual and the possession, in higher animals, of an elaborate defence against infection. It may be helpful to summarize the responses of the latter to virus infections. Broadly speaking the diverse cells of the immune system provide the immune response (Fig. 12.1) and any other infected cells make the anti-viral protein type 1 interferon.

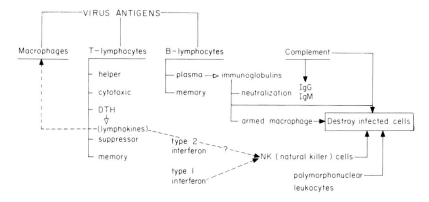

Figure 12.1 Summary of responses of the immune system and associated cells to viruses and virus-infected cells.

THE IMMUNE SYSTEM

There is antibody-mediated immunity and cell-mediated immunity. Antibodies (immunoglobulins) are made by plasma cells. Their immediate progenitors are B-lymphocytes which react with foreign molecules (antigens) and are stimulated to divide and differentiate to form a large number of plasma cells and memory cells. The plasma cells synthesize and secrete free molecules of antibody which react specifically with the stimulating antigen. Memory cells have the same specificity but do not synthesize antibody. Because many are made as a result of the initial reaction with antigen, memory cells enable an animal to give an enhanced response to antigen when it is encountered for a second time and they then develop into plasma cells.

Immunoglobulins (Ig) have the basic pattern seen in Fig. 12.2. The five types are distinguished by having different heavy chains called α, γ, δ, ε and μ which characterize IgA, IgG, IgD, IgE and IgM respectively.

Immunoglobulins act directly against viruses by attaching to and neutralizing viruses (Chapter 4). Alternatively they can become attached (IgG only) by the

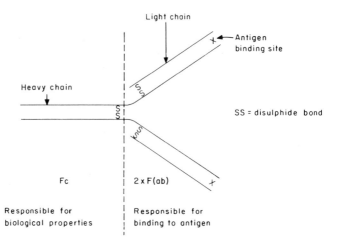

Figure 12.2   Generalized immunoglobulin molecule.

F_c region to macrophages which enables these cells to use the specificity of antibody to recognize viral antigens exposed on the surface of infected cells. Finally IgG and IgM can activate complement, a cascade enzyme system; when activated this releases components which attract immune cells to the location of the antigen and expresses a phospholipase activity which lyses cells bearing the antigen or enveloped viruses.

The other arm of the immune system consists of non-immunoglobulin secreting cells which are developed from T-lymphocytes. These have receptors for antigen, probably immunoglobulin light chain molecules, in their plasma membranes and are stimulated into cell division and differentiation by contact with antigen. Some of these daughter cells remain as memory cells, the others have a variety of functions in immunity. Briefly the major types of differentiated T cells are:

1.   Helper T cells which assist B-lymphocytes to react with some antigens and without which these antibodies are not made.
2.   Cytotoxic T cells which directly destroy cells carrying foreign (e.g. viral) antigens.
3.   Cells responsible for delayed-type hypersensitivity. These secrete lymphokines which are soluble molecules having a wide variety of specific effects on other cells. The well known macrophage inhibitory factor (MIF) is one of these. Lymphokines recruit cells to the site of antigenic stimulation and alter the permeability of blood vessels in that area. Hence one observes a swelling and reddening at that site due to an influx of fluid and cells.
4.   Suppressor T cells which depress the immune response. These have a role in controlling the immune responsiveness of other cells.

170

Associated cells

These cells are not derived from B- or T-lymphocytes. The best known is the macrophage which can engulf (phagocytose) particles of foreign material. Unlike B- or T-lymphocytes its activity is not confined to a specific antigen. However macrophages have receptors for the F_c region of IgG molecules and such 'armed' macrophages have antibody specificity conferred upon them. Macrophages are found mainly in the body cavities, but another family of phagocytes, the polymorphonuclear leukocytes, occur in the blood stream. Polymorphs are not armed with antibody. K (killer) cells do not phagocytose but cause contact lysis of cells to which IgG has bound.

The last cells to have been recognized in playing a role against virus infections are the natural killer (NK) cells. These can arise early, within 2 days of infection, and their activity is stimulated by type 1 interferon. NK cells are not armed with antibody either.

INTERFERON

The interferons, first discovered by Isaacs in 1957, probably hold the greatest promise for anti-viral chemotherapy. Isaacs and his collaborator had incubated the chorio-allantoic membrane from embryonated chicken eggs in a suspension of heat-killed influenza virus and then transferred this membrane to buffer. After storage for 24 hours the membranes were discarded and the buffer tested for anti-viral activity. This was done by placing a fresh piece of membrane in the buffer and then challenging the membrane with active influenza virus. It was found that membranes so treated did not support the growth of active virus in contrast to untreated membranes. It was concluded that an extracellular product had been liberated in response to viral infection and this substance was named *interferon*.

We now know that there are two types of interferon. Isaacs discovered type 1 interferon while type 2 is released by stimulation of certain T cells by the appropriate antigen, i.e. it is a lymphokine.

Type 1 interferons

These have been extensively purified from several sources. Interferons are proteins which have a carbohydrate group essential for their activity as this is destroyed by agents which remove the carbohydrate moiety of glycoproteins. The anti-viral activity can be sensitively measured by inhibition of the incorporation of radioactive uridine into viral RNA in cells infected by a togavirus. Activity is expressed as the amount of interferon required to reduce the normal level of viral RNA by 50% and this is arbitrarily defined as one unit. Purified interferon has a high activity of around 10^9 units per mg of protein (which is of the same order as the hormones).

A great variety of viruses can induce interferons and once induced this interferon is active against the whole spectrum of viruses, not just the virus responsible for induction. However, interferon appears to be more specific for cells of the species in which it was induced since interferon obtained from mice affords little protection to rat, hamster, chicken or monkey cells.

The mode of action of interferon can be divided into two stages (Fig. 12.3): *induction* which results in the release of interferon, and production of an *anti-viral state* in other cells by the released interferon. Induction of interferon in human cells is controlled by chromosome 9. All multiplying viruses induce interferon and it is believed that double-stranded (ds) RNA is the specific inducer. If this is so then it means that even DNA viruses must synthesize RNA molecules with some double-stranded regions.

Free interferon initiates an anti-viral state in other cells by binding to a receptor on their cell surface. Chromosome 21 is required and probably codes for the synthesis of receptor molecules. The anti-viral state has two modes of action which are initiated in cells by the combination of ds RNA in the presence of interferon. Firstly, dsRNA stimulates the phosphorylation of certain proteins, which impairs their function in the initiation of protein synthesis. Secondly, dsRNA activates a ribonuclease which degrades mRNA and hence also stops protein synthesis. In fact, the dsRNA directly stimulates the synthesis of 'two-five A' (a trinucleotide pppA2'p5'A2'p5'A), which, in turn, activates the ribonuclease.

Finally, the question of specificity. How does interferon prevent only viral protein synthesis? In some cases it does not, and virus multiplication is prevented by death of the cell. Otherwise the cell survives, presumably because the inhibitory mechanisms discriminate between viral and host protein synthesis. However we do not understand how this operates.

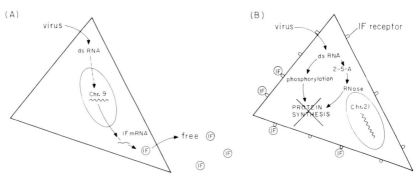

Figure 12.3 (A) Induction of type 1 interferon (IF) by viral double-stranded (ds) RNA. The dashed line indicates that we do not know if dsRNA is required to enter the nucleus to initiate interferon mRNA synthesis or whether it acts through an intermediate. (B) Establishing of an anti-viral state through the inhibition of protein synthesis.

Immune (type 2) interferon

When they are infected, cells of the immune system not only synthesize type 1 interferon like any other cells, but also release type 2 interferon upon reaction with viral (and other) antigens. This can be demonstrated by infecting an animal with a particular virus and then some days later reacting its T cells *in vitro* with the same virus. Macrophages and probably B cells also respond in a similar way.

Type 2 interferon has anti-viral properties like type 1 but is antigenically different and less stable at pH 2. Its induction is not dependent upon the presence of dsRNA and it appears to supplement the type 1 interferon response. Type 2 interferon is of advantage to the host during infections by viruses which switch off host RNA and protein synthesis before type 1 can be made and may have a role in stimulating parts of the immune response, such as NK cells.

13 Virus-host interactions

It is useful to remember that viruses are parasites and that biological success of a virus depends absolutely upon the success of the host species. Hence the strategy of a virus in nature must take into account that it is a disadvantage to kill the host or impair its reproductive ability. In this chapter we shall be concerned with viruses of eukaryotes and their host cells, drawing largely upon the animal kingdom for examples. We shall lay particular emphasis upon the complexity of these interactions and the multitude of factors which share responsibility for the final outcome of infection. These studies represent one of the most important frontiers of modern virology and probably the first without a precedent in the bacteriophage systems. Consequently plant and animal virologists must go it alone!

A classification of the various types of virus-host interactions is given in Table 13.1 and each category is dealt with in turn later. However this is only intended as a guide as there is really a spectrum of virus-host interactions and the divisions are imposed for convenience. It will also become apparent that a single virus may appear in several of the categories depending on the nature of its interaction with the host.

Table 13.1 A classification of virus infections at the level of the whole organism

	Infectious progeny	Cell death (lysis)	Symptoms
Acute	+	+	+
Inapparent	+	+	−
Chronic	+	+	+ or −
Persistent	≪+	−	−
Latent	−	−	−
Slowly progressive	+	+	eventually +
Tumourigenic: productive	+	−	+
:defective	−	−	+

ACUTE INFECTIONS

Acute infections are the nearest by analogy to the classical one-step growth curve of the T-even bacteriophages, except of course that the time scale is measured in days rather than minutes. Infection of organisms can be described in terms of the symptoms produced and by a variety of laboratory tests. Without the latter no identification of the causative agent is complete. Laboratory tests include isolation and titration of infectious virus, detection of viral antigens in the blood by immunoelectrophoresis, detection of viral antigens by fluorescent antibody staining in cells obtained by biopsy, and the direct identification of virus using the electron microscope possibly in conjunction with antibody which

174

will agglutinate homologous virus particles. In an acute infection, infectious progeny are produced and infected cells die. Further cycles of multiplication ensue and eventually the first symptoms appear. Thus the infection has been progressing for several days before we are even aware of the fact (Fig. 13.1 and particularly 13.2). Fortunately most people recover from virus infections and it is appropriate at this point to consider which mechanisms are responsible for allowing us to survive our first encounter with a particular virus, i.e. a primary infection. It is difficult to obtain unequivocal data from such a complex situation

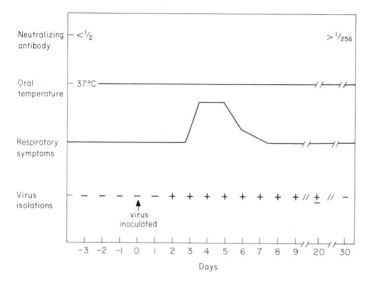

Figure 13.1 Virus infections are measured both by symptoms and by laboratory test for the presence of virus. The figure illustrates a 'common cold' in man caused by intranasal inoculation of a rhinovirus.

far removed from the study of cells in culture, and the only major point to emerge is that it is impossible to generalize about recovery mechanisms to date since different viruses are susceptible to different parts of the immune system (Table 13.2). However it is likely that interferon, which is synthesized as soon as multiplication commences, plays a major role in limiting the spread of infection. Recovery is assisted by the destruction of infected cells, which have viral antigens on their surface, by cytotoxic T cells. Non-specific killer cells appear very shortly after infection and they too directly destroy the infected cells (see Fig. 12.1). We appear to suffer no harm from losing a number of infected cells in this way, probably because they represent a very small proportion of the total present in a tissue. At the end of an infection, free virus is removed by both specific antibody and macrophages.

Recovery may be summarized in military terms:

1.   Stand by to repel attack (interferon).
2.   Search and destroy, counter-attack (cytotoxic T cells; antibody + complement;

175

NK cells; armed macrophages and killer cells).

3. Mopping up (macrophages; antibody).

A well-studied example of an acute infection is the disease caused in mice by mouse pox (ectromelia) virus (Poxviridae; Class I) (Fig. 13.2). To obtain such

Table 13.2 Recovery from primary viral infections

Immunity responsible for recovery	Virus Group
T-lymphocyte derived cells	Herpes
	Pox
Antibody from B-lymphocyte derived (plasma) cells	Picorna
	Orthomyxo
	Paramyxo
	Toga

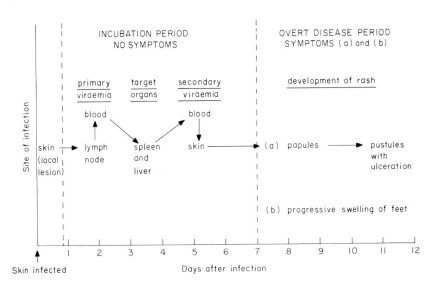

Figure 13.2 The course of an acute infection: mouse pox virus in mice.

data mice are infected and killed at intervals and various organs are dissected out and ground up so that the amount of virus there can be determined. It is important to remember that in all infections, extensive multiplication takes place during the symptomless incubation period; that any multicellular organism consists of many types of differentiated cells organized into tissues which go towards forming a highly integrated system of organs—and that viruses home in and multiply in certain 'target' organs; and finally that expression of the disease is the outcome of the combined properties of the virus and its host.

176

INAPPARENT INFECTIONS

These are the commonest infections and as their name implies there are no symptoms. In all other regards these are the same as acute infections. Evidence of infection comes only from laboratory isolation of the virus or by a subsequent rise in a specific antibody. A virus which causes an inapparent infection has evolved to a favourable equilibrium with its host: the enteroviruses (Class IV, Picornaviridae) which multiply in the gut are one such group. The classic example is poliovirus which causes no symptoms in over 90% of infections. The inapparent infection is clearly the result of interaction between the virus and its *natural* host since there are many examples of such a virus causing lethal infections when it infects a different host, e.g. yellow fever virus (Class IV, Togaviridae) causes an inapparent infection of Old World monkeys but results in a severe infection in man, and is fatal in some New World monkeys. Inapparent infections have the same duration and are cleared by the same means as acute infections.

The following chronic, persistent and latent infections have in common the feature that bodily defences cannot get rid of the viruses responsible. We shall discuss the various ways in which this may come about.

CHRONIC INFECTIONS

We have defined a chronic infection in Table 13.1 as if it were an acute or inapparent infection which is not terminated by defence processes. The main difference from persistent or latent infections is that large amounts of infectious virus are produced throughout. The pertinent question is why the combination of interferon and immune response is ineffective. In general, it appears that the infection has a deleterious effect upon the immune response so that virus multiplication continues (Table 13.3). At the moment it is not known why

Table 13.3 Possible ways in which viruses depress immune responsiveness and establish chronic infections.

Tolerance resulting from infection *in utero* or of neonates	
Immunosuppression through virus	a) multiplying in lymphocytes
	b) inhibiting proliferation of lymphocytes
	c) stimulating suppressor T cells
Infection of macrophages	

interferon is ineffective in curtailing infection. For interferon to be effective it must be present at the right place, at the right time and in a sufficiently high concentration. Perhaps viruses can in some way upset this formula.

Classic examples of chronic infections occur when lymphocytic choriomeningitis virus (Class V, Arenaviridae) or Aleutian disease virus (Class II, Parvoviridae) are inoculated into new-born mice or hamsters respectively. The animals are immunologically tolerant to the viral antigens and virus can be found in large quantities in the circulation and tissues. One of the unexplained features is that tolerance is rarely complete but presumably the reduced immune response is unable to cope with the infection. Similarly, little interferon is made although the viruses are sensitive to its action. Neonatally infected animals remain healthy whereas animals infected as adults mount an immune response to the virus which has the fatal effect of 'immunological suicide'.

In man, a chronic infection results from neonatal contact with rubella virus (Class IV, Togaviridae). Another outstanding example occurs when adults are infected with hepatitis B virus (an unclassified DNA virus), which is the causative agent of serum hepatitis. After an initial acute infection virus may continue to be present for a lifetime. Effects on T-cell responses are not yet known, but circulating antibody against viral antigens is present and the deposition of the resulting antigen-antibody complexes may lead to an immune complex disease. Hence the appearance of symptoms in this instance depends not on the cytolytic effects of virus but upon the relative proportions of viral antigens and antibody.

PERSISTENT INFECTIONS AND LATENT INFECTIONS

Although these categories are formally separated in Table 13.1, in practice they are difficult to tell apart. Unlike the persistent SV5 infection of cells in culture (p. 165) very little virus can be detected in animals with persistent infections. The Herpetoviridae (Class I) are a diverse and ubiquitous group which normally cause persistent infections although they may start as, or develop into, an acute infection. Persistence is maintained by a balance between viral properties and the host's defence mechanisms. Breakdown of persistence leads to the activation of an acute infection which occurs frequently when people are immunosuppressed and an apparently healthy person then becomes acutely ill. Natural activation is seen with herpes simplex (HSV) virus type 1 and with varicella-zoster virus (Herpetoviridae) (Fig. 13.3). After the initial acute infection the virus is maintained

	Acute infection	→ Latent/persistent infection	+ immunosuppression or 'physiological factors'	→ Acute infection
HSV 1	Fever blisters or cold sores around the mouth	In dorsal root ganglion of trigeminal nerve	e.g. Strong sunlight, menstruation, tension	Fever blisters or cold sores
VZV	Chicken pox	In dorsal root ganglia	'Ageing'	Shingles/Zoster

Figure 13.3 The origin of persistent/latent infections in an individual and the breakdown of persistence/latency with herpes simplex type 1 virus (HSV 1) or varicella-zoster virus (VZV).

in the dorsal root ganglia of spinal or cranial nerves supplying the original area of infection. Upon activation the virus descends the sensory nerve and erupts into an acute infection at the nerve ending in the skin. HSV 1 infection is maintained in the presence of large amounts of antibody and a T cell response both of which probably contribute to keeping the virus in its persistent/latent state. HSV 1 has been isolated by co-cultivation of the trigeminal ganglia with cells from a sensitive line showing either that small amounts of infectious virus are produced *in vivo* (a persistent infection), or that virus is activated under the conditions of culture (a latent infection).

178

SLOWLY PROGRESSIVE DISEASES

The agents which are presumed responsible have been misnamed 'slow viruses'. Many affect the central nervous system and one of the best known is the scrapie agent of sheep which causes the animal to scrape or scratch itself against obstacles. The causative agent has not been isolated hence it cannot be positively identified as a virus but it has some properties similar to viroids.

The relationship between multiplication of the scrapie agent in the mouse and disease is illustrated in Fig. 13.4. Scrapie multiplies initially outside the central

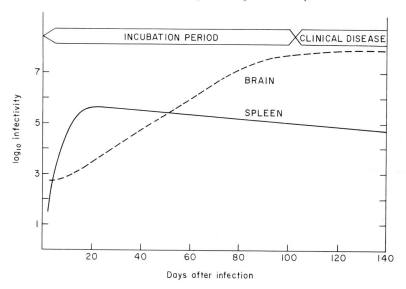

Figure 13.4 Multiplication of the scrapie agent in mice and the time course of the disease.

nervous system in organs such as spleen or lymph nodes. Peak titres are reached in the spleen in about 20 days which is a rate comparable to some acute virus infections. Infectivity accumulates in the brain over a longer period and reaches a higher concentration than elsewhere. Clinical disease becomes apparent when brain infectivity attains a plateau at 3-4 months after infection. This is the best of scrapie experimental systems, for clinical disease in its natural host does not occur for about 2 years. If it is true that the scrapie agent codes for no proteins, there are no antigens against which the immune system can respond. Hence there is the unique situation where the immune response plays no part in the manifestation of the disease.

The slowly progressive diseases have similarities initially with inapparent infections and later with persistent infections but the late development of the disease symptoms gives them a unique category in Table 13.1.

VIRUS-INDUCED TUMOURS

All viruses which cause tumours have DNA as their genetic material but some, the RNA tumour viruses, have RNA in their virions, i.e. they belong to Class VI.

DNA 'tumour' viruses normally cause an acute infection and it is rare that a tumour results, e.g. most young people are infected by EB virus (Class 1, Herpetoviridae) which causes infectious mononucleosis or glandular fever, yet occurrence of tumours (Burkitt lymphoma, nasopharyngeal carcinoma) with which the virus has been associated is rare. Demonstration of tumourigenicity in the laboratory is frequently made of necessity under 'unnatural' circumstances which are known to be favourable to the development of the tumours. Important factors are the genetic attributes and age of the infected animal: certain inbred lines develop tumours more readily than others and young animals are more susceptible because their immune system is not fully mature. If no tumours result it may be necessary to transfer cells transformed by the virus *in vitro* into the animal to demonstrate viral tumourigenicity.

The diagnostic criterion for a tumour virus in the laboratory is the production of a transformed cell from a normal cell. This involves many complex changes in cellular properties which will be discussed more fully in Chapter 14. However not all transformed cells are tumourigenic when transplanted to appropriate animals. So it would seem that another step in addition to transformation is needed to produce a cancer cell. Thus transforming viruses which cause cells to divide more rapidly than normal are at one end of a spectrum occupied at the other by those causing cellular destruction. Tumours fall into two groups: those which produce infectious virus and those which do not. The latter are the more common: they contain viruses which are defective and unable to multiply and those in which multiplication is in some way repressed (like lysogeny). The methods described above for detecting viral components in latently infected cells may be used successfully on tumour cells. However, it is suspected that some tumours are caused by 'hit and run' viruses which cannot be detected by present methodology. This suspicion was strengthened by finding that cells transformed by adenoviruses in culture contained as little as 7% of the genome of the original infecting virus. It may be that even smaller pieces of genome are all that is required to maintain the transformed state.

Like other virus-host interactions which have been discussed above, the induction of tumours depends upon the balance of a complex situation. This can be viewed as shown in Fig. 13.5 where initially infection may be acute or in a persistent or latent form. The transformation event may be enhanced by or require additional factors (such as chemical carcinogens in Fig. 13.5), but these

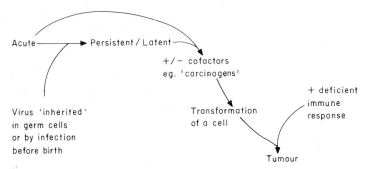

Figure 13.5 Possible relationship between virus infection and tumourigenesis.

could be broadly regarded as the physiological state of the cell. How often transformation of cells takes place in the animal we do not know but it is likely that this is frequent and that on most occasions the immune system recognizes and destroys the transformed cell. Tumour formation probably requires that the immune system is in some way deficient, a state perhaps induced by infection or by ageing.

INTERACTIONS WITH VIRUSES WHICH TAKE PLACE OUTSIDE THE HOST

In this section we shall continue with virus-host interactions by considering how viruses are transmitted from an infected to a susceptible host. Since most viruses are inherently unstable and lose infectivity rapidly, this is an important step in their life cycle. Knowledge of how viruses are transmitted may enable us to break the cycle at this stage and thus prevent further infections. We shall deal with transmission of infection by the respiratory route, by the faecal-oral route, by the urino-genital tract and by 'mechanical' means.

Transmission via the respiratory tract

Not only viruses which cause respiratory infections but also some viruses causing generalized infections such as measles and smallpox are contracted by this route. After virus has multiplied, it either re-infects host cells or escapes from the respiratory tract in the aerosol which results from our normal activities such as talking, coughing, sneezing (and particularly singing!). These aerosols are inhaled and give rise to a 'droplet infection' of a susceptible individual. The size of droplets is important as those of large ($>10\,\mu$m) diameter rapidly fall to the ground and the smallest ($<0.3\,\mu$m) dry very quickly resulting in accelerated inactivation of virus contained therein. Thus the middle-sized range of droplets are those which transmit infection. The size of these will determine where the droplets are entrapped by respiratory system of the recipient, since the 'baffles' lining the nasal cavities remove the larger airborne particles while the smaller may penetrate deep into the lung.

It seems that the increase in nasal secretions which accompany many respiratory infections favour the dispersal of the viruses responsible and the increase in coughing and sneezing increases the production of infected aerosols. However, transmission experiments from people infected with a rhinovirus to susceptibles sitting opposite at a table proved singularly unsuccessful. This has led to the suggestion that only particular individuals may shed sufficient virus, produce excess nasal secretion and/or aerosols containing optimum sized particles and act as efficient spreaders of infection.

Linked with transmission of respiratory infections are environmental factors which result in seasonal variation in the amount of illness or frequency of isolation of a virus. Influenza, for example, is well known as a winter disease. One can point to variations in both the environment (e.g. temperature and humidity) and in social behaviour, (crowding together in winter with poor ventilation) which could well affect virus transmission and hence the seasonal incidence of virus diseases, but a full explanation of these complex phenomena is not yet available. Since the virus can survive for only a limited time outside the cell, it is fair to assume that infected individuals are present continuously in the population.

Transmission by faecal-oral route

At some stage of infection all these viruses multiply in, or are passed into, the alimentary tract. One example is poliovirus and there are many similar entero-viruses spread in an identical manner; another is infectious hepatitis A virus (Picornaviridae, Class IV). All these viruses infect after being ingested so that their spread is favoured by poor sanitation and poor personal hygiene. Not surprisingly, when one remembers how small children investigate strange objects by putting them in their mouth, many of these viruses cause infections in early childhood.

One would expect to see a reduction of enteric viruses when sanitation is improved and basically this has been the experience. However there have been some unpleasant surprises. In conditions of poor sanitation, poliovirus infects young children and results usually in the appearance of an inapparent gut infection. With improved sanitation poliovirus is not contracted until adolescence, and associated with this shift in age distribution is an increase in the incidence of paralytic poliomyelitis. This is another example of increased severity of a disease seen when a virus infects an 'un-natural' host and in this case age is the important difference.

Transmission via the urino-genital tract

Some viruses are excreted in urine and may re-infect as described above. Of greater importance are those which are spread by sexual contact or are transmitted to the offspring by the mother. Herpes simplex type 2 is spread by sexual contact and has been cited as a possible cause of cervical cancer. Serum hepatitis B virus is also spread by this route probably through abrasions in the skin and mucous membranes.

The transmission of virus from mother to offspring is called 'vertical' transmission in contrast to the 'horizontal' transmission between other individuals. Rubella virus (Togaviridae) is transmitted vertically and is responsible for congenital malformation if the fetus is infected in the first three months of life. It is impossible to decide with certainty whether a virus is transmitted to the zygote from an infected oocyte or sperm, or whether the zygote is infected from virus present in cells of the uterus, or via the bloodstream in placental mammals. Any virus which infects the mother while she is producing eggs or offspring can in theory be vertically transmitted, while those which have genomes integrated with that of the host cannot help but be transmitted by this route. Infection can also result from passage of the ovum down an infected oviduct or by contact with any infected cells on its way to the exterior. Although it does not belong in this section, another means of transmission of certain viruses to the newborn animal occurs via infected milk.

Transmission by 'mechanical' means

Under this umbrella are included most plant and some animal viruses which do not infect their hosts simply by virus-cell contact but require some extra agency to permit infection. Plant viruses are transmitted into plant cells by animals (vectors), such as aphids, leaf hoppers, beetles and nematodes, which feed on plants, or by non-specific abrasions made to the plant tissue which expose cells sufficiently for them to be infected. Transmission by plant-feeding animals is a specific process and only certain species are implicated in the transmission of a particular virus (see Chapter 18). Some viruses multiply in the vector while others are passively transmitted.

Animal viruses which are spread by mosquitoes are found predominantly in the alpha- and flavi-viruses (Togaviridae) and the Bunyaviridae. As with the plant viruses transmission is limited to certain species and it is known that the viruses cited multiply in the insect vector as well as the vertebrate host. Other arthropod-borne animal viruses do not multiply in their vector, an example being myxoma virus which is spread between rabbits by passive transfer on the mouth parts of infected mosquitoes in Australia and rabbit fleas in the United Kingdom.

RECAPITULATION

It is worth reiterating here the major features of virus-host interactions:

1. The majority of infections result in a satisfactory end for both the host population (if not the individual) and the virus. In other words infections are dealt with by the host's defence processes, so that the host recovers and can reproduce itself but not before the virus has time to multiply and perpetuate itself by being transmitted to other susceptible individuals.
2. Infections which kill (or reproductively incapacitate) large sections of a population are rare. When they occur one should suspect that there has been a change in the genetic compatibility of host and virus, i.e. the 'wrong' host is infected. Although we are far from understanding the nature of this genetic compatibility, it is probably the product of long-term evolution.

14 Vaccines and chemotherapy: the prevention and treatment of virus diseases

Man is concerned particularly to protect himself, his plants and his animals against death and economic loss caused by virus infections. In animals (including man) immunization with vaccines has been far more effective than chemotherapy (see also page 74) but for many viruses the vaccine is far from ideal. These problems will be discussed below. The rationale of immunization is to raise a specific immunity without that individual experiencing the disease. Since plants do not have an immune system, prevention of plant virus diseases has relied upon other means such as breeding of plants which are genetically resistant to the virus or its vector, or by control of the vector. Control of animal virus diseases through improvements in public health and personal hygiene is another important aspect of prevention.

PRINCIPAL REQUIREMENTS OF A VACCINE

A vaccine may be either infectious ('live') or non-infectious ('killed'). Upon administration, all vaccines should have the properties outlined in the following subsections.

Cause less severe disease than the natural infection

In other words a vaccine causing illness can be tolerated when the alternative is more serious illness or death, but obviously vaccine manufacturers strive for a product causing the minimum of discomfort. The process of producing a virus strain which causes a reduced amount of disease is called *attenuation*. The disease-causing virus is referred to as the *virulent* strain and the attenuated strain as *avirulent*.

Avirulent strains are usually obtained by selecting and enriching for such variants which occur naturally. This is achieved empirically and experience has shown that it can be helped by multiplication in cells unrelated to those of the normal host, multiplication at sub-physiological temperature or by recombination with an avirulent laboratory strain. The use of strains which are antigenically related to the virulent strain and cause less disease in that host has long been known since Jenner in 1798 used cowpox virus to vaccinate against smallpox. To this day we use a descendant of cowpox virus, vaccinia, for the same purpose and the term 'vaccine' has become synonymous with any immunogen used against an infectious disease. Naturally enough, killed vaccines should cause no disease at all. However, since a virulent strain is being used it is essential to kill *every* infectious particle present (see p. 186). An advantage of killed preparations is that other unknown contaminating viruses may also be killed by the same process. However, a severe disadvantage is that the killed virus does not multiply so that an immunizing dose has to contain far more virus than a dose of live vaccine. This increases the cost and the amount of impurities present which may result in hypersensitivity reactions when these substances are experienced again.

Effective and long-lasting immunity

The reasons for these properties are scientific and sociological. As explained in Chapter 13, different viruses are susceptible to different parts of the immune system hence it is necessary for the vaccine to stimulate immunity of the correct type and in the correct location for it to be effective. This is a problem with the killed influenza virus vaccines currently in use which are administered by injection and do not raise IgA antibody at the respiratory epithelial surface. However sufficient IgG to be protective is thought to diffuse from the blood stream to the respiratory surface in about 50% of individuals immunized. If an immunity is not effective then the virus being immunized against may still be able to multiply. This occurs with foot-and-mouth disease virus of cattle and the partial immunity provides a selection pressure which favours the multiplication of new antigenic variants of the virus which in time replace the pre-existing strain. Even animals with effective immunity against the original strain will not be protected from the new strain hence the long, expensive process of vaccine development and immunization would have to be repeated.

Public faith in any preventative medicine is rapidly lost if that measure is not effective in the majority of cases. Consequently, to avoid bad publicity a vaccine must protect the majority of individuals who have received it. Immunization should result in long-lasting immunity as it is surprisingly difficult to persuade people to come forward to be immunized. Hence a 'single-shot' vaccine which requires only one visit to the doctor is the ideal. Some employers have circumvented this problem by offering immunization at the place of work. In tropical countries where immunization is often most needed, it is particularly important that a vaccine has a good shelf-life (i.e. that its infectivity or immunogenicity is stable) or immunization will be ineffective.

Genetic stability

An attenuated live virus vaccine must not revert to virulence when it multiplies in the immunized individual. Vaccines which are formed by 'killing' virulent strains of viruses must not be restored to infectivity by genetic interactions with viruses occurring naturally in the recipient.

ADVANTAGES, DISADVANTAGES AND DIFFICULTIES ASSOCIATED WITH LIVE AND KILLED VACCINES

Inactivation

A killed vaccine must have no infectivity and yet be sufficiently immunogenic to provoke a protective immunity. Inactivating agents must ideally inactivate viral nucleic acid function and not merely attack the outer virion proteins since infectious nucleic acid could be released by the action of the cell on the coat. Secondly, inactivating agents must lend themselves to the industrial scale of manufacturing processes. Formaldehyde and β-propiolactone are two which have been used. The former cross-links proteins through ε-amino groups of lysine residues and reacts with the amino groups of nucleotides. β-propiolactone inactivates viruses by alkylation of nucleic acids and proteins. The basic problem of preparing a killed vaccine is that every one of the 1 000 000 000 or so infectious particles which are contained in an immunizing dose of virus must be rendered non-infectious and this was the problem which faced Jonas Salk before the first poliovirus vaccine could be presented to the general public in 1953. Inactivation of the infectivity of polio- and other viruses is exponential and has the kinetics shown in Fig. 14.1. Frequently it is found that a small fraction of the population

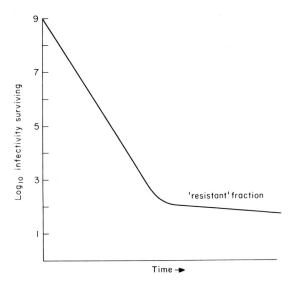

Figure 14.1 Exponential kinetics of virus inactivation. Note the 'resistant' fraction which is inactivated more slowly.

is inactivated far more slowly than the majority so that the whole virus population has to be kept in contact with the inactivating agent for much longer than is predicted by the initial rate of inactivation. This has the concomitant danger that immunogenicity will be destroyed. Examination of the 'resistant' fraction shows that it seems to result from inefficient inactivation of particles trapped in aggregates or clumps. Live vaccines present none of these problems but the difficulties of producing a genetically stable attenuated strain are at least as great. For poliovirus these were overcome by Albert Sabin who had to attenuate all three serotypes to produce the successful live vaccine which has been in use since 1957.

Routes of administration

The unavoidable difficulty of killed vaccines is that they do not multiply! Hence they do not reach and stimulate immunity in those areas of the body where infectious virus would normally be found. However, providing that the killed vaccine is a potent immunogen, the high levels of serum IgG which result from infection of the vaccine can often serve in place of local immunity presumably because there is a sufficient concentration to diffuse to the extremities.

The means of administration is an important factor in persuading people to accept a vaccine. Injections are painful and unpopular, particularly with small children and their mothers. A compressed air device which is less painful and quicker, has been used for mass immunization in the U.S.A. in place of the traditional hypodermic syringe and needle. A disadvantage of killed viruses is that two injections are required to build up a sufficient secondary immune response. This could in theory be avoided by emulsifying vaccine with an adjuvant. When injected this remains as a lump and releases immunogen over a long period and there is the additional advantage that the adjuvant also stimulates the activity of the immune system. However adjuvants cause local irritation and may be carcinogenic hence none has been deemed satisfactory for use in man although they are used in other animals.

Cost

The production of vaccines for use in man is subject to legislation by the Food and Drug Administration in the U.S.A. and by the Department of Health in the U.K. The precautions laid down for the production of a safe vaccine are necessarily stringent. However this means that the cost of the final product will be correspondingly high. Precautions increase as knowledge is obtained about hitherto undetected viruses in cells used in vaccine manufacture, or other potential hazards, and as they do so the costs increase yet again. Regulations governing the use of veterinary vaccines are not so demanding and the product is correspondingly cheaper. The other major aspect which directly affects the cost is the quantity of vaccine which has to be produced. Live vaccines are cheap to produce compared with killed vaccines, since they only have to initiate an infection whereas an immunizing dose of killed vaccine has to be provided in two injections. Finally, the costs involved with packaging, distribution and administration to the recipient are considerable. These all add up to make vaccination a luxury which poor countries, usually those in the greatest need, cannot afford. Provision of vaccines is one of the ways in which the richer countries can and often do provide valuable aid.

Multiple vaccination

The immune system can respond to more than one antigen at a time hence it is possible to immunize with a vaccine 'cocktail'. This has advantages in minimizing the number of injections and inconvenience. In practice, satisfactory immunity results from multiple killed vaccines, but problems with multiple live vaccines may arise due to mutual interference in multiplication possibly as a result of the induction of interferon (see later).

'Unusual' situations

Listed below are some of the circumstances when the normal immunization procedure is inadvisable or even dangerous. In all these situations only a killed vaccine should be used.

1. Very young children sometimes develop a more severe infection than occurs in an older child.
2. Certain clinical conditions can make an infection more severe than normal, e.g. people with eczema are prone to a generalized infection with vaccinia virus instead of the local infection at the site of introduction of the vaccine.
3. Fetal development can be deranged by virus infection, e.g. by rubella virus.
4. A prior virus infection can interfere with the multiplication of a live vaccine with the result that immunity is not established.

Under certain circumstances it has been found that the acquisition of immunity as a result of administration of a killed vaccine potentiated the disease when the wild virus was contracted. This occurred with respiratory synoytial virus (Paramyxoviridae; Class V) which causes a lower respiratory infection in young children. Increased severity of disease is thought to have resulted from IgG, provoked by the vaccine, forming immune complexes in the respiratory tract which in turn stimulated an influx of fluid and immune cells to that site and impeded breathing. The danger of this situation is that it is not revealed until the vaccine is tested in man. In the example cited the disease was potentiated only in infants, the group which was in greatest need of protection, and this demonstrates a need to test a vaccine in all age groups. Immunization with respiratory syncytial virus must now await the development of a live vaccine.

It is too late to immunize when symptoms of an infection are already apparent. However, beneficial results can sometimes be achieved by giving immunoglobulin with activity against the infecting virus and of course it helps to have identified the causative virus. However this passive immunity lasts only as long as the immunoglobulin survives in the body. A further complication is that one must be prepared to treat an anaphylactic response if successive doses of this 'foreign' immunoglobulin are needed. Other emergency measures not involving the immune system involve chemotherapy including treatment with interferon (see following section).

CHEMOTHERAPY OF VIRUS INFECTIONS

Ever since the successful introduction of antibiotics to control bacterial infections there has been the hope that similar treatments for virus infections were just around the corner. Except for the use of chemicals in certain specific circumstances this hope has not yet been realised. The reason is that the virus multiplication is tied so intimately to the cell that chemicals cannot discriminate between them. However, viruses do have unique features so, in theory, chemicals specific for these should be able to serve as effective chemotherapeutic agents. A chemical would be effective if it blocked or interfered with any of the stages of multiplication, i.e. with attachment, replication, transcription, translation, assembly or release of progeny particles.

Chemicals are most widely used to treat DNA virus infections of the conjunctiva and cornea of the eye. The agents used are halogenated pyrimidines which are incorporated into viral DNA and prevent replication and transcription. These components are incorporated into cellular DNA hence their use is restricted to the poorly vascularized surface of the eye. In practice 0.1% 5'-iodo-2'-deoxy-uridine, an analogue of thymidine, improved healing in 72% infections with herpes simplex type 1, vaccinia and adenoviruses. In extreme circumstances such as in cases of encephalitis caused by herpes virus type 1 or vaccinia viruses, treatment with analogues of cytidine (cytosine arabinoside) or adenosine (adenine arabinoside) may be used. These chemicals are extremely poisonous and can only be used where death is the likely alternative. Another anti-viral compound is amantadine (p. 74) which has been used prophylactically and therapeutically to treat influenza virus infections. In an epidemic the number of cases in the population under study fell from 14.1% to 3.6% with amantadine prophylaxis. Methisazone (N-methylisatin β-thiosemicarbazone) has been used to treat smallpox infection. Not an ideal anti-viral compound, it is very insoluble, a strong emetic and is rapidly degraded in the body. As a consequence evaluation has been difficult since the amount of methisazone remaining in the body is hard to determine. Fortunately the W.H.O. campaign to eradicate smallpox by vaccination has pre-empted these problems.

Interferon therapy

Interferon's properties of universal anti-viral activity and high specific activity coupled with its lack of toxicity make it potentially the ideal therapeutic agent. Much effort has been expended to obtain it in sufficient quantity and at reasonable cost for clinical use. This problem has been partially resolved by inducing human leucocytes, obtained from blood banks, to make interferon. Still greater quantities would be needed to put interferon to general use. Attempts to persuade cells to increase their interferon yield or to isolate the human interferon gene and to introduce it into a bacterium are being made but it is too early to judge if these genetic manipulations will be successful. Why are such vast quantities of interferon needed when its specific activity is so high? The reasons are that interferon is unstable in the body and is rapidly degraded with a half-life of about 3 hours and also that it is not possible to apply interferon solely to those cells in which virus is multiplying. Some of the high estimates of interferon needed for anti-viral activity may have resulted from the experience of attempts to treat respiratory infections since there are suggestions that generalized infections may respond to lower doses.

One of the first successful experiments in interferon therapy was at the Common Cold Research Unit at Salisbury, U.K. where volunteers were given

Table 14.1 Treatment with interferon of a common cold caused by a rhinovirus infection

	Colds	Virus Isolations
Interferon treated	0/16	3/16
Mock treated	5/16	13/16

14×10^6 units of interferon as a nasal spray 4 times per day including the day before infection with a rhinovirus (Table 14.1). Recently more encouraging results have been achieved by treating people who have chronic serum hepatitis B virus infections. This is a transmissible infection which is debilitating and sometimes fatal. Viral antigens and DNA polymerase present in the blood of patients decreased transiently in response to large doses of interferon (up to 10^7 units/day), but there was a sustained decrease in response to continued low doses (10^5-10^6 units/day) given over a period of 4 months which extended beyond the period of treatment. Undoubtedly there is optimism amongst interferon workers today.

15 Tumour viruses

The great majority of viruses of vertebrates are not oncogenic[†]: that is to say, they do not have the ability to initiate a tumour. However, the normal manifestation of infection by some deoxyriboviruses is a benign tumour and several other deoxyriboviruses can produce malignant tumours when injected into newborn animals. Only a small number of riboviruses have been unequivocally associated with neoplastic disease but these are particularly interesting since they are oncogenic under natural conditions.

As well as causing tumours, both groups of viruses are able to transform mammalian cells in tissue culture. Transformed cells are recognized by their ability to grow in low concentrations of serum and in soft agar, their loss of contact inhibition (cf. page 4), the loss of existing antigens and by the presence of new ones. Transformed cells may form tumours in animals but since tumour formation involves complex interactions with the immunological system of the host animal, transformation in culture is not an infallible indicator of malignancy *in vivo*. Transformation permits the study of early events required for the development of malignant potential by infected cells and opens up the possibility of studying the acquisition of new genetic material by mammalian cells.

THE ONCOGENIC DEOXYRIBOVIRUSES

Members of three families of viruses, the Papovaviridae, the Adenoviridae and the Herpetoviridae, are known to have oncogenic potential. Numerous papovaviruses cause warts (papillomas) in man and his domestic animals. Another causes no demonstrable disease in laboratory mice but when artificially inoculated into infant rodents causes a wide variety of tumours, hence its name 'poly-oma'. Yet another, simian virus 40 (SV40), came to light during the preparation of poliovirus vaccine in monkey kidney cells. SV40 is apt to grow out from apparently normal monkey kidney cells and since it produces vacuoles in infected cells it is sometimes called vacuolating agent. The name of this group: papovavirus, is thus derived from the initials of its chief members: PA for papilloma, PO for polyoma and VA for vacuolating agent. Adenoviruses were first isolated from cultivated cells from adenoids and tonsils. Shortly afterwards they were found to be responsible for an outbreak of sore throat amongst military recruits. The prefix adeno- given to those viruses indicates an affinity for adenoid or glandular tissues. In 1962 it was shown that human adenovirus type 12 produces sarcomas when inoculated into newborn animals and subsequent work has shown that several other adenoviruses of human and animal origin are also oncogenic for newborn, but not adult, rodents. Three herpesviruses are suspected of being oncogenic. These are Epstein-Barr virus found in African children suffering

[†] A glossary of terms is provided at the end of the chapter.

from Burkitts lymphoma, the virus associated with neurolymphomatosis of chickens (Marek's disease) and the virus associated with the Lucké renal carcinoma of leopard frogs. Of these viruses, the best studied are polyoma and SV40 and we shall restrict our subsequent discussion almost entirely to these two viruses.

Cells in culture can respond to infection by polyoma or SV40 in one of two ways: the virus may undergo a productive infection (in a 'permissive' cell) or an abortive infection (in a 'non-permissive' cell). During an abortive infection the virus may enter the cell and cause the synthesis of some virus-specified products without multiplying. Whether the response of a cell to infection will be productive or abortive depends ultimately on the type of cell. Thus mouse cells are permissive for polyoma, and hamster and rat cells non-permissive. SV40, on the other hand, productively infects monkey and human cells but abortively infects mouse, hamster, rat and rabbit cells.

The properties of transformed cells

A variable, but very low fraction of abortively infected cells is transformed but cells which are subject to productive infection are only rarely transformed. Cells which have been transformed are readily detected by their altered growth characteristics, as outlined above, and by the ease with which they are agglutinated by certain plant proteins such as wheat germ agglutinin and concanavalin A, which bind strongly to glycoproteins and are collectively known as 'lectins'.

If untransformed cells are exposed to mild doses of trypsin, the cells undergo a further round of cell division transiently rendering them more susceptible to agglutination by lectins. This observation suggests that transformation results in rearrangement of sites which bind glycoproteins causing the cells to lose the ability to regulate their growth and to respond to contact inhibition. This idea demands that the surface of an untransformed cell changes transiently at each mitosis in such a way that glycoprotein sites are exposed. Thus the cell would be programmed to undergo a further division unless overriding signals prevented this action. Transformation, by fixing the cell surface permanently in the state which is only transiently assumed at mitosis, would cause the cell to divide indefinitely.

Unfortunately, for the above hypothesis, there is no really significant difference between the amounts of radioactively labelled concanavalin A or wheat germ agglutinin bound by a range of untransformed cells, or cells transformed with polyoma and SV40. However, the transformed cells are agglutinated by smaller amounts of these lectins than are the untransformed cells. Thus there is no simple relationship between the amount of lectin which a cell binds and the ease with which it can be agglutinated. The difference in response to lectins still remains a valuable parameter of transformed cells but it must reflect the topographical distribution of binding sites rather than their number.

The events leading to transformation

Under the proper conditions, cultures of mouse cells may be maintained in the non-dividing state for long periods of time, the cells being arrested in the G1 period of the growth cycle. If such a culture is infected with a high concentration of SV40, cells are induced to synthesize DNA, to divide and to assume a transformed appearance. However, the majority of these cells are abortively transformed since after a few cell divisions they revert to their previous non-transformed appearance. A few of the colonies remain permanently transformed and we can speculate that only in these is the viral DNA integrated with that of the host cell. The number of divisions necessary to permit the development of colonies of transformed cells has been determined by infecting a non-dividing culture and replating cells at various dilutions, so that the cells completed a known number of cell divisions before confluence was again obtained. To determine whether that fraction of cells which are induced to replicate their DNA after SV40 infection also give rise to stable transformants, confluent monolayers of non-permissive cells were infected with a high MOI of SV40. The induction of DNA synthesis was monitored with a suitable radioactive label and the dividing cells separated from the non-dividing cells by velocity sedimentation. Up to 60% of the cells were induced to synthesize DNA within 24 hours of infection. Once separated, the transformation frequencies of the two populations were measured and, as expected, a much higher proportion (55%) of dividing cells than non-dividing cells were transformed.

Apparently the establishment of transformation requires some virus-cell interaction which is not expressed when the cells are resting in the G1 phase of the cell cycle but which is expressed either during S phase (DNA replication) or the G2 phase. Since transformation is blocked by adding interferon before S, some virus gene function is evidently required in addition to a cell function and/or division.

Why cell division is necessary for the stabilization of transformation remains an enigma. It may be that the viral genome can only integrate when the host cell's DNA replicates. During evolution SV40 and polyoma have probably acquired the capacity to induce the replication of the DNA of the cells they infect because these viruses rely on cellular DNA replication for the replication of their own genomes. Because many cells in an adult animal are not dividing rapidly, a viral gene which can induce a cell to divide would greatly increase the range of cell types in which the virus can replicate.

In addition to being readily agglutinated by lectins, cells transformed by polyoma virus acquire three new tumour antigens called (big) T, (middle) t and (little) t which are distinct from the structural proteins of the virus. These antigens are identical with proteins produced in the early stages of productive infection although the relative amounts may vary. Immunofluorescence studies show that T-antigen occurs in the nucleus where it binds to DNA, middle t is located in the plasma membrane and in the cytoplasm and little t is cytoplasmic. A polyoma virus TSTA (tumour specific transplantation antigen) occurs in the plasma membrane of transformed cells and is recognized by transplant rejection tests in which cell-mediated immunity is predominant. The relationship between the three tumour antigens and TSTA is not clear but its location in the plasma membrane makes middle t a likely candidate. There is an analogous situation in SV40-transformed cells which has TSTA, T-, little t- (but no middle t-) antigens and an additional nuclear antigen, the U (for uncharacterized) antigen. All are immunologically distinct from polyoma virus antigens. Initiation of transformation by SV40 requires the production of both T and little t antigens but only little t is necessary for maintaining the transformed state as certain lines of SV40-transformed cells have lost the T-antigen and remain transformed. A paradox of this system is that SV40 TSTA is related to the nuclear T-antigen.

The stable transformation of mouse cells by SV40 depends on the cells dividing at least once after infection.

The events leading to transformation

Under the proper conditions, cultures of mouse cells may be maintained in the non-dividing state for long periods of time, the cells being arrested in the G1 period of the growth cycle. If such a culture is infected with a high concentration of SV40, cells are induced to synthesize DNA, to divide and to assume a transformed appearance. However, the majority of these cells are abortively transformed since after a few cell divisions they revert to their previous non-transformed appearance. A few of the colonies remain permanently transformed and we can speculate that only in these is the viral DNA integrated with that of the host cell. The number of divisions necessary to permit the development of colonies of transformed cells has been determined by infecting a non-dividing culture and replating cells at various dilutions, so that the cells completed a known number of cell divisions before confluence was again obtained. To determine whether that fraction of cells which are induced to replicate their DNA after SV40 infection also give rise to stable transformants, confluent monolayers of non-permissive cells were infected with a high MOI of SV40. The induction of DNA synthesis was monitored with a suitable radioactive label and the dividing cells separated from the non-dividing cells by velocity sedimentation. Up to 60% of the cells were induced to synthesize DNA within 24 hours of infection. Once separated, the transformation frequencies of the two populations were measured and, as expected, a much higher proportion (55%) of dividing cells than non-dividing cells were transformed.

Apparently the establishment of transformation requires some virus-cell interaction which is not expressed when the cells are resting in the G1 phase of the cell cycle but which is expressed either during S phase (DNA replication) or the G2 phase. Since transformation is blocked by adding interferon before S, some virus gene function is evidently required in addition to a cell function and/or division.

Why cell division is necessary for the stabilization of transformation remains an enigma. It may be that the viral genome can only integrate when the host cell's DNA replicates. During evolution SV40 and polyoma have probably acquired the capacity to induce the replication of the DNA of the cells they infect because these viruses rely on cellular DNA replication for the replication of their own genomes. Because many cells in an adult animal are not dividing rapidly, a viral gene which can induce a cell to divide would greatly increase the range of cell types in which the virus can replicate.

The state of the DNA in transformed cells

Viral DNA must be present in transformed cells since DNA extracted from such cells hybridizes specifically with RNA made *in vitro* using purified viral DNA as template. This DNA must be in close association with the host cell chromosomes since it co-sediments with nuclear DNA in alkaline sucrose gradients. The polyoma and SV40 genomes are circular and possibly they are integrated into the chromosome in a manner analogous to that whereby *E. coli* becomes lysogenized by phage λ (page 151). Cells transformed by SV40 contain a complete copy of the viral genome since they yield infectious virus by co-cultivation or fusion with permissive cells and the extracted DNA is infectious.

Transcription of the viral genome in transformed cells

During a productive infection about one half of the SV40 genome is transcribed early in infection, before the replication of viral DNA begins, and comprises the class of early RNA. After DNA replication has begun the remainder of the genome is transcribed so that late in infection both early and late transcripts are made (see also Fig. 8.9).

Transcription of SV40 DNA *in vitro* by *E. coli* RNA polymerase results in an RNA product (cRNA) which hybridizes almost exclusively with one strand of the viral DNA (Fig. 15.1). This cRNA fails to anneal with early viral RNA but can anneal with late RNA species. Thus early viral RNA must be transcribed *in vivo* from the same strand that is used as a template *in vitro* by *E. coli* RNA

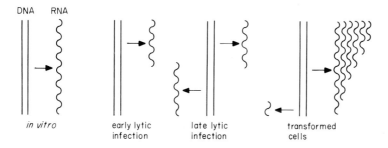

Figure 15.1 Transcription by SV40 under various conditions.

polymerase. The appearance of late RNA will depend on a change in the strand transcribed, a phenomenon already described for phages λ and T4. As yet there is no explanation why DNA replication must occur before there is a switch to late transcription. Only a fragment of late RNA is found in cells transformed by SV40. In fact, the majority of the RNA synthesized in transformed cells appears to be identical to that synthesized early in a productive infection. What happens to prevent the synthesis of late RNA? During the process of inserting the viral genetic material into the mammalian cell genome, the circular SV40 DNA is probably converted to a linear molecule. If the interruption of the circular form occurs inside the 'late' region close to the late promoter, only early RNA could be synthesized. However, this hypothesis demands firstly that viral and host transcription should have the same orientation, and secondly that the early and late genes are not intermingled—a fact which has now been shown experimentally.

Polyoma virus contains only 3×10^6 daltons of DNA which is sufficient to code for about 200 000 daltons of protein, about eight normal-sized proteins. On the face of things it should be straightforward to define such a simple genome, including those genes responsible for the maintenance and establishment of transformation, by isolating conditional lethal mutants. Indeed, such mutants have been identified. The *ts*a mutant of polyoma carries a temperature-sensitive lesion in a gene for the establishment but not the maintenance of transformation; *ts*3 mutants carry a *ts* lesion in a gene for the maintenance of transformation. Hamster cells transformed by *ts*3 and maintained at the permissive temperature, 32°C, have all the characteristics of cells transformed by wild-type polyoma. On shifting to 39°C, *ts*3 transformed cells become sensitive to contact inhibition, i.e. they lose their transformed character at the restrictive temperature. By contrast, cells transformed by wild-type virus are insensitive to contact inhibition at both temperatures.

The *ts*3 mutation also affects the response of permissive cells supporting the replication of the virus. Usually when polyoma virus replicates in mouse cells, host DNA synthesis is induced concomitant with the replication of the viral genome and the cells are induced to divide. At 32°C the *ts*3 mutant behaves in these respects like wild-type virus but at 39°C both the induction of cell division and cell DNA synthesis by *ts*3 are impaired. Most likely these are pleiotropic effects resulting from a change in the chemistry of the cell surface.

Similarities between transformation and lysogeny

There is a clear analogy between the virus-transformed cancer cell and the bacterium lysogenized by a temperate phage. In both cases the cell is not destroyed and produces no virus. Lysogenized bacteria can often, but not always, be induced by physical and chemical treatment. In a similar fashion, SV40 transformed cells are known to liberate virus either spontaneously, as happens on rare occasions, or as a result of cell fusion. If monkey cells, which support cytocidal infection with SV40, are fused with SV40-transformed cells, induction of SV40 occurs in about 10% of the resulting heterokaryons. The knowledge that phage DNA is integrated into the host chromosome when a bacterium is lysogenized has always coloured ideas about how eukaryotic cells are transformed by papovaviruses. Indeed, this analogy inspired the experiments which established that SV40 DNA becomes stably associated with the DNA of the transformed cell. However, this analogy should not be taken too far for the factors which determine whether a cell is a target for transformation differ radically from those which determine whether a bacterial cell is lysogenized or is instead lysed by an infecting phage. Lysogeny is due to the production of a repressor but all the evidence suggests that this has no counterpart in transformed cells.

One interesting feature of papovavirus replication is the production of pseudovirions which are viral capsids containing fragments of host DNA about the size of the viral genome. When these were first discovered there was hope that they could be used for transduction of animal cells. Unfortunately, this does not seem likely. Since the genome of the papovavirus is only 3×10^6 daltons and that of the host about 10^{12} daltons there is not much chance of a particular gene being included in a pseudovirion, even allowing for considerable gene duplication in eukaryotic cells.

Transformation by other DNA tumour viruses

Much less is known about the transformation of cells by adenoviruses. As with the papovaviruses, each adenovirus-transformed cell conatins adenovirus-specific T and transplantation antigens and the equivalent of several molecules of viral DNA sequences integrated into the chromosomes. However these are incomplete genomes and infectious virus cannot be recovered from transformed cells. Although only about 0.02% of the DNA from malignant cells will anneal specifically with DNA extracted from purified adenovirus, this viral DNA must be preferentially transcribed since 2% of the RNA recovered from the polysomes is viral specific. Hybridization experiments show that this viral RNA is transcribed from less than 10% of the viral genome.

Transformation by adenoviruses does not require integration of even the majority of the genome but only of a specific part of the DNA. All adenovirus-transformed cells contain a region from one end of the genome (known by convention as the left-hand part) which represents 10% of the total genomic DNA. In other experiments cells could be transformed by fragments of isolated DNA of 1×10^6 mol. wt. (equivalent to about 5% of the genome) or by a 1.6×10^6 mol. wt. fragment derived specifically by restriction endonuclease cleavage from the left-hand end of the genome.

Interest is now focused on how such a small fragment of viral DNA, which can code for no more than two proteins of 50 000 mol. wt., produces the extensive cellular changes which are associated with transformation. Maybe adenoviruses synthesize an equivalent of the single putative transformation protein which has been isolated recently from cells infected with an RNA tumour virus (see below).

ONCOGENIC RIBOVIRUSES

Only a small number of riboviruses are known to be oncogenic and all belong to the genus Oncovirinae of the Retroviridae (formerly the leukoviruses). These viruses are enveloped and have, as their genetic material, a single-stranded RNA molecule of mol. wt. 3×10^6. This RNA is associated with proteins which are antigenically identical in all oncornaviruses infecting the same host, i.e. they are group specific antigens. The group can be divided into two types: B-type[†] viruses which cause breast tumours, and perhaps other carcinomas, and C-type[†] viruses which cause leukaemias and sarcomas. A-type virus particles are always intra-cellular and are believed to be precursors of B- and C-type particles.

Some strains of mouse mammary tumour virus, which is a typical carcinoma-

---

[†] These are trivial names used to describe two different kinds of particles seen in the electron microscope.

197

inducing B-type virus, are transmitted vertically. That is to say, they can be transmitted from parent to offspring by crossing the placenta or by suckling infected milk. This virus thus provides a system for studying the induction of carcinomas which are, of course, the most prevalent cancers of man. Until recently, all attempts to infect cells *in vitro* failed but recent success opens the way for further investigations which will complement those of infected cell lines that can readily be established from mammary carcinoma tissue. Such cell lines continue to produce virus which is capable of infecting mice but not laboratory cell lines. It seems that this virus is fastidious about the type and state of differentiation of its host cells and their hormonal environment. Clearly, the new *in vitro* cell system will aid progress towards the unravelling of the biology of this virus.

The C-type viruses can infect cells in tissue culture and consequently much more is known about them. Although the avian oncornaviruses have been the most intensely studied, they suffer the disadvantage that permanent chicken cell lines for their growth and maintenance are not available. Permanent cell lines for the growth of mammalian C-type viruses are available but one disadvantage of such immortal lines is that they no longer bear much resemblance to their cell of origin. The most obvious indicators of these changes are that the cells are heteroploid and have dedifferentiated so that they no longer possess the specialized morphology and biochemical abilities that they possessed *in vivo*.

If fibroblast cells are infected with either avian (ALV) or murine leukaemia virus (MLV), very little morphological change occurs. The cells are neither transformed nor lysed making a focal assay impossible. In order to study the replication of some murine viruses the XC test has recently been developed. If mouse cells which are replicating leukaemia virus are mixed with XC rat tumour cells, syncytia are formed. These syncytia appear as plaques in the layer of XC cells and can be used to quantitate the virus. ALV and MLV cause leukaemia when injected into the appropriate animal, and transform epithelial but not fibroblast cells *in vitro*.

The behaviour of murine sarcoma virus (MSV) is quite the opposite since it cannot replicate on its own but is capable of transforming fibroblasts (and not epithelial cells) *in vitro*. Replicating stocks of the virus can be obtained but these contain MLV which provides a necessary helper function. If a stock of MSV is allowed to infect mouse cells, and foci are picked, two types of foci may be detected; one class produces two viruses, an XC plaque-forming virus (i.e. a leukaemia virus) and a focus-forming virus (i.e. a sarcoma virus). However there are also foci which are the result of the sarcoma virus infection alone. These comprise cells which produce no infectious leukaemia or sarcoma virus and are therefore known as 'non-producer' cells. However, if leukaemia virus is added to such cells, the sarcoma virus genome can be rescued. Some strains of Rous sarcoma virus (which infects chickens) behave in a similar fashion but the situation is complex and the reader is referred to specialized texts for details.

The unexpected involvement of DNA in viral RNA replication

Early studies showed that DNA seems to play a critical role in the multiplication of RNA tumour viruses and in their ability to transform. Infection and trans-

198

formation by these viruses can be prevented by inhibitors of DNA synthesis added during the first 8-12 hours after exposure of the cells to the virus. Also, the formation of virions is sensitive to actinomycin D suggesting a requirement for DNA-dependent RNA synthesis. These data led Howard Temin of the University of Wisconsin to propose his 'provirus' theory which postulated the transfer of the information of the infecting RNA to a DNA copy which then serves as a template for the synthesis of viral RNA.

Temin's theory required the presence in infected cells of a unique enzyme, an RNA-dependent DNA polymerase or 'reverse transcriptase'. Until 1970, no enzyme had been found in any type of cell which could synthesize DNA from an RNA template. It was clear that if such an enzyme existed then the RNA tumour virus must induce its synthesis soon after infection or else carry the enzyme into the cell as part of the virion. Since there already was precedence for the occurrence of polymerases in animal viruses (page 102) a search was begun for a reverse transcriptase in several oncornaviruses. Simultaneously in 1970, David Baltimore of MIT and Temin reported the presence of such an enzyme. Purified virions were disrupted with detergent and incubated with dATP, dCTP, dGTP and ^{3}H-dTTP. This resulted in a rapid incorporation of label into acid-insoluble material, i.e. DNA. By omitting one of the deoxyribonucleotide triphosphates or by pretreating the virus with ribonuclease this incorporation could be prevented proving that the enzyme was indeed making DNA from an RNA template.

Properties of oncornavirus genome

Much recent genetical, biochemical and biophysical work has contributed to understanding the organization of the genomes of oncornaviruses. It became apparent that the early estimates of coding capacity were falsely high when it was discovered that each virus contained at least two copies of the viral genome.

Early studies showed that genomic RNA sedimented at 70S which is equivalent to a mol. wt. of around 7×10^6. However, when denatured, this RNA sedimented at 35S, indicating that there were two molecules of equal size. Did these contain different information or were they identical? This question was resolved when a method was devised to map oligonucleotides derived by specific RNase T_1 digestion by separating them in 2-dimensions. T_1 cuts RNA leaving guanylate at the new 3' terminus and the number of large oligonucleotides (runs of nucleotides not containing an internal guanosine residue) gives a measure of size or complexity of the RNA, since the runs are more likely to occur in large molecules. Analysis of oncornavirus RNA gave a complexity expected for a molecule of around 3×10^6 mol. wt. Thus each virion contains two copies of the RNA and is sometimes called diploid. Electron microscopy has confirmed that the two RNA molecules are linked together as shown in Fig. 15.2. These RNA molecules are associated together by their 5' terminal regions and this structure may be necessary for replication (see p. 202). It may seem surprising to recall that oncornaviruses have a genome approximately the same size as poliovirus.

Chemical characterization of the viral genome has shown that it resembles a typical eukaryotic mRNA having messenger polarity, a poly A tract at the 3'

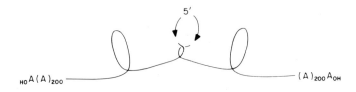

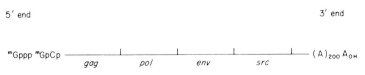

Figure 15.2 Top: structural map of the oncornavirus genome showing identical molecules held together at their 5' ends. Each has an intramolecular loop. Bottom: genetic map of Rous sarcoma virus (not to scale).

terminus and a cap structure at the 5' terminus. The order and number of the four major genes has now been determined for Rous sarcoma virus (RSV). These are known by the abbreviations *gag*, coding for proteins which include the *g*roup specific *a*ntigen of the nucleocapsid; *pol*, coding for the virion-associated *pol*ymerase or reverse transcriptase; *env*, coding for the virion *env*elope glycoprotein, and *src* the *s*a*rc*oma or transforming gene (Fig. 15.2). Mapping of the genome was achieved in part by using an *src* deletion mutant which does not transform cells. This mutant is analogous to MSV (above) in having a smaller genome than the fully competent parental virus (MLV and RSV respectively).

Far from being poorly understood, as much is now known about the strategy of the genome of oncornaviruses as any other virus.

Properties of reverse transcriptase

The template requirements of reverse transcriptase have been investigated. Exogenous nucleic acid, when added to disrupted virus could also serve as template, often giving much higher rates of synthesis than occur on the endogenous 60-70S genomic RNA. Polyribonucleotides acted as efficient templates although the enzyme from different viruses showed slightly different specificities for the four homoribopolymers. Synthesis of the complementary polydeoxyribonucleotides continued until an amount of polymer equal to the amount of initial template had been synthesized. However, all the ribopolymers were poor templates unless a complementary oligodeoxyribonucleotide, which could be as short as four bases, was added as a primer. Even with a primer, polydeoxyribonucleotides were much poorer templates than the corresponding homologous polyribonucleotide and in most cases yielded no detectable synthesis.

Is a primer required to initiate reverse transcription of the tumour virus genome? When the product of the reaction of endogenous viral RNA with reverse transcriptase is analysed, label appears in a molecule which has the properties of an RNA-DNA hybrid. Thus the possibility arises that the primer for DNA synthesis is a short RNA molecule. If the nascent DNA chain is covalently linked to a primer RNA then the product at early times should be an RNA chain with a few deoxynucleotides attached. Such a molecule will have a density close to that of single-stranded RNA. As the reaction proceeds and the nascent DNA chain is elongated while the length of the RNA primer remains unchanged the density of the product should decrease progressively until it approaches that of pure single-stranded DNA. The behaviour of the products, made at various times by incubating purified avian myeloblastosis virus particles in a polymerase assay system, fulfil these predictions exactly when they are banded in Cs_2SO_4 gradients. Since the RNA component can be removed by alkali and ribonuclease, but not by denaturation, it must be covalently linked to the DNA.

What is the nature of the endogenous primer RNA? In the *in vitro* reaction this is known to be a cellular 4S tRNA. There is one molecule per viral genome. The tRNA is intimately associated near the 5′ terminus but can be separated after denaturation. Strains of oncornaviruses contain only one type of tRNA, e.g. Rous sarcoma virus has tryptophan tRNA and Moloney murine leukaemia virus has proline tRNA. It is not known how these tRNAs are specifically selected.

Physically, the reverse transcriptase protein of avian strains is composed of two sub-units: α (60 000 mol. wt.) which has enzymatic activities and β (90 000

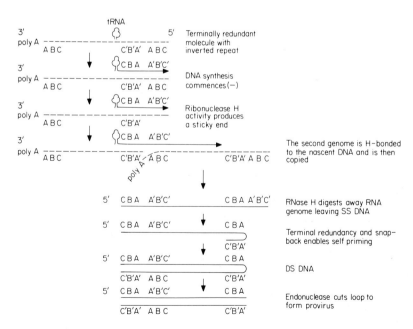

Figure 15.3 Hypothetical model to show the functions of reverse transcriptase during transcription of an oncornavirus genome into the DNA provirus.

201

mol. wt.) which binds to the tRNA primer and is enzymatically inactive. The **α** sub-unit has 3 activities which (a) transcribe genomic RNA into DNA, (b) transcribe DNA into DNA, and (c) digest the RNA strand from RNA:DNA hybrids (ribonuclease H).

The *in vitro* product of the reverse transcription of MLV can be separated into two fractions by sedimentation in sucrose gradients. These two fractions were analysed for their content of single- and double-stranded DNA and DNA-RNA hybrid by digestion with nucleases of known specificity and by isopycnic centrifugation in Cs_2SO_4. The major fraction early in the reaction contained equal amounts of single-stranded DNA and hybrid but little double-stranded DNA. After extensive synthesis the major fraction contained equal amounts of single-stranded and double-stranded DNA but little hybrid. In the presence of actinomycin D the predominant product was single-stranded DNA, a small proportion of which was of genome size.

The basic problem is to explain how RNA is copied into DNA when synthesis commences with the tRNA primer at the 5′ terminus of the virion RNA (remembering that nucleic acid synthesis always proceeds from 3′ → 5′ terminus of the template). It would, of course, have been easier to explain if nature had put the tRNA primer at the 3′ end! However Fig. 15.3 presents a model which suggests how DNA may be synthesized and at the same time explains why the viruses are diploid.

Integration and replication of oncornaviruses

How does the DNA copy of the oncornavirus genome become integrated with the

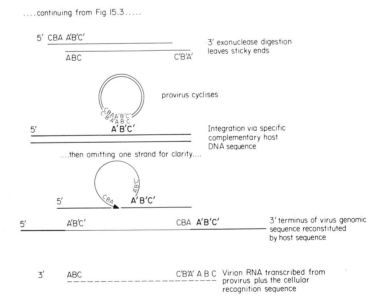

Figure 15.4 Hypothetical model for the specific integration of the DNA copy of an oncornavirus genome and its subsequent transcription into virion RNA.

202

host's DNA? The model in Fig. 15.3 can be extended to explain these phenomena as shown in Fig. 15.4. A cyclized form of the provirus has been found experimentally in infected cells by density gradient centrifugation in caesium chloride containing ethidium bromide (p. 52, Chapter 3) but the mechanism of integration is an ingenious hypothesis originating from W. A. Haseltine and D. Baltimore. It is proposed that the linear provirus is cyclized by the formation of sticky ends after digestion with endonuclease. However, this means that each strand now lacks the terminal 3′ sequence of the viral genome. The model goes on to propose that there is one (or more) specific recognition sites and that these are identical with the missing 3′ sequence. Integration follows recognition and thus the full genomic sequence is reconstituted and complete virion RNA can be transcribed.

Is reverse transcriptase necessary for transformation?

Does the infection and transformation of cells by RNA tumour viruses depend absolutely upon the activity of reverse transcriptase? The enzyme has been found in all C-type RNA containing tumour viruses as well as in murine and monkey mammary tumour viruses, which are B-type viruses. It has not been found in a variety of other RNA viruses which replicate by budding from membranes suggesting that the enzyme was not just adventitiously adsorbed from the cell. The most satisfactory evidence, however, is the observation that mutants of RSV with a temperature sensitive reverse transcriptase cannot initiate infection at the restrictive temperature. The enzyme has also been found in two other genera of the Retroviridae which contain visna and maedi and the foamy viruses. Visna virus, found in sheep, causes a progressive neurologic disease not unlike multiple sclerosis in man. It is known that it can exist in a latent state and that its replication in cell culture is also affected by actinomycin D. Thus, investigators pursued the parallel to the oncoviruses and found a similar RNA dependent DNA polymerase. Recently it has been shown that it can transform mouse cells but whether it causes tumours is not known. The other group of viruses in which reverse transcriptase has been found is a group of syncytium-forming viruses which are generally referred to as 'foamy' viruses. They have not been shown to be oncogenic.

Thus the number of classes of viruses containing the enzyme is currently larger than the group of known oncogenic viruses. In addition, there are reports that the enzyme has been found in uninfected cells. Absolute proof of this, however, is difficult since it is almost impossible to rule out the possibility of 'silent' infection with other tumour viruses. Also, the template used in screening uninfected cells is important since some of these can also be copied by host cell DNA polymerase. Fortunately the viral enzyme can be separated from host cell polymerase and in addition it prefers single-stranded RNA as template.

If reverse transcriptase is involved in transformation then it might be possible to isolate proviral DNA from transformed cells and use it to infect and transform other cells. To examine this possibility DNA was extracted from hamster cells infected with a temperature-sensitive mutant of RSV. After purification the DNA was used to infect chick fibroblasts which were then examined for foci of transformed cells. Foci were found which were temperature sensitive as expected.

It seems clear, therefore, that transformed cells do indeed contain at least one DNA provirus which can be isolated in an infectious state.

Various rifamycins (inhibitors of transcription) have been examined for their effect on reverse transcriptase *in vitro*. These derivatives fall into three groups: those which fail to inhibit (Class A), those which partially inhibit (Class B) and those which are potent inhibitors (Class C). *In vivo*, Class C derivatives block transformation completely whereas Class A reduce it by 25% and Class B by 50%. Moreover, a correlation between inhibition of transcriptase and loss of infectivity was not found when similar experiments were repeated with non-oncogenic riboviruses which contain RNA-dependent RNA polymerase instead of DNA-dependent RNA polymerase.

Finally, a strain of RSV has been isolated, the RSV (O) strain, which unaided is unable to transform and which lacks reverse transcriptase. All these data strongly suggest that reverse transcriptase is involved in the infection and transformation of cells by the oncogenic riboviruses.

The process of transformation

Transformation by RNA or by DNA viruses is thought of either as the result of the action of gene products or of the integration event(s) *per se*. In the case of oncornaviruses there would be a single transforming RNA or protein, the putative product of the *src* gene. It is a problem to conceive how a single gene product could be the cause of the extensive and varied changes comprising transformation unless it was able to have an effect on a variety of cellular genes or their products. Recently several research groups claim to have isolated the *src* protein which has a mol. wt. of 60 000, turns over rapidly and has protein kinase activity (i.e. it can phosphorylate other proteins—and this is a common means of metabolic control in eukaryotic cells). The existence of such a transformation protein would make it easy to explain the fact that certain *ts* mutants of RSV give rise to cells having a transformed morphology at the permissive temperature but when those cells are raised to the restrictive temperature the morphology reverts to normal.

The possible cellular changes brought about by the integration of the *src* gene are discussed below in the context of the 'oncogene theory'.

Origin of RNA tumour viruses

Several theories have been advanced to explain the origin of RNA tumour viruses but it should be borne in mind that different viruses may have had different origins and so all theories may be correct (or, indeed, incorrect!).

Provirus theory

This suggests that tumours are the symptoms of infections caused by catching viruses in the same way as, for example, influenza (i.e. tumours result from a horizontal infection). Temin proposed that the DNA provirus was the intermediate both in establishing the tumour and in the production of progeny virions.

Certainly the experimental RSV and MLV/MSV systems behave like this but other mechanisms are possible.

Oncogene theory

This states that the majority of vertebrate cellular genomes contain DNA copies of oncornavirus genes and that these are inherited vertically like any cellular gene. Infection as described by the provirus theory may have been responsible originally for establishing the oncogene early on in evolution of the species.

The viral gene remains unexpressed under the action of cellular repressors until an external process (possibly chemical, irradiation or viral) derepresses the system and the oncogene and perhaps other viral genes are expressed. In this category are the large number of endogenous viruses shown to be present in non-malignant tissue.

Protovirus theory

This was advanced by Temin to explain the occurrence of spontaneous tumours. The theory suggests that normal cells exchange information in the form of RNA during development and that cellular reverse transcriptase activity incorporates this information into DNA. By some random event one cell accumulates sufficient cellular genetic information to give rise to an oncornavirus *de novo* (this would need information for reverse transcriptase, structural proteins and transformation). The new virus would then infect surrounding cells and result in the formation of a tumour.

Evolution of the *sarc* gene

Recent hybridization studies have shown that sequences called *sarc*, which are related to the *src* gene, can be detected in *all* cells of *all* avian species tested (Table 15.1). *sarc* is apparently not part of the genome of an endogenous virus since RAV_0 (the endogenous virus) is integrated in a different chromosome. The

Table 15.1 Evolution of *sarc* sequences in avian DNA (adapted from Stehelin *et al.*, 1976, *Nature* **260**, p. 172)

DNA	% annealing	Tm (°C) of annealed $cDNA_{src}$*	Phylogenetic distance from chicken (years × 10^6)
XC cells infected with:			
RSV	56	81	-
Chicken	52	77	0
Turkey	30	72	40
Duck	16	71	80
Emu	15	70	100

*^{3}H DNA complementary to viral *src* was annealed with cold DNA from the sources indicated.

efficiency with which cellular DNA annealed with the *src* gene from RSV and the thermal stability of such hybrids indicates the degree of relatedness and this corresponds to the evolution of birds as deduced from fossil evidence and other sources. It is concluded that *sarc* was present in the ancestral bird and has evolved at the average rate for avian genes.

Why have cells conserved what is apparently a harmful gene? A clue to the answer to this problem was the finding that the *sarc* genes are transcribed at a low level in normal cells, and this suggested that *sarc* may have a function which is universally required by normal cells. How can this be reconciled with the viral *src* causing transformation? This apparently conflicting situation could be resolved by the suggestion that the normal cellular function is the regulation of cell multiplication while infection with an oncornavirus containing *src* might result in an increase in the *sarc* and *src* product sufficient to alter the balance of cellular behaviour in favour of transformation.

Hybridization studies in mammals have shown that sequences present in the genomes of an endogenous virus of cats (RD114) and of baboons are conserved in other feline and simian species respectively in proportion to their known phylogenetic relationships. The study of the significance of oncornavirus genomes to normal development is obviously of utmost interest.

THE SEARCH FOR HUMAN CANCER VIRUSES

Trivial benign tumours, such as warts, are most certainly caused by viruses but as yet we do not know if any human malignancies are of viral origin. It would be surprising if viruses were not responsible for some human cancers since they induce different types of malignancy in several species of animal. Definitive proof of the viral aetiology of human cancer will only be obtained if Koch's postulates can be fulfilled. Thus it must be shown that the suspected virus is encountered in every case of the disease and that pure preparations of the virus cause the disease when injected into healthy animals. Clearly, this latter test is out of the question for agents suspected of causing cancer and other, less direct criteria will have to be fulfilled.

Numerous research groups have spent a considerable amount of time (and money!) attempting to isolate viruses from human cancers. The first report of the isolation of a putative human cancer virus was the discovery of Epstein-Barr (EB) virus, a herpesvirus, in tissue cultures derived from patients suffering from Burkitt's lymphoma. However, EB virus is the cause of glandular fever, an almost universal human infection, which after the acute stage results in a typical herpesvirus persistent infection of lymphocytes. Thus it is not unexpected to find EB virus in the lymphoma. Whether or not EB virus is the causative agent of Burkitt's lymphoma is an open question even though it can transform human cells in culture. Analogous arguments also face the association of herpesvirus type 2 with cervical carcinoma: did the virus cause the cancer or merely invade the malignant cells after they had appeared?

In 1971, much excitement was generated by the discovery of a C-type virus, RD114 virus, in human cells. A line of human rhabdomyosarcoma cells (RD cells), in which there was no evidence for the production of any RNA virus, had been inoculated into fetal kittens. Four of these kittens developed sarcomas, and cells from these tumours (RD114 cells) were found to be producing a C-type virus. The producer cells were obviously the progeny of the initial human cell inoculum since they had a human karyotype but was the virus of human origin? From the outset a human origin was favoured because the internal proteins of the virus, i.e. the group-specific antigens and the reverse transcriptase, were not immunologically related to those of the feline leukaemia viruses. Furthermore, RD114 virus could grow in human but not in feline cells. However, attempts to isolate similar viruses from the original RD cells were always unsuccessful and no evidence of a viral genome related to RD114 could be found in these cells using molecular hybridization. Then it was shown that a virus identical with RD114 could be spontaneously produced from a line of feline kidney fibroblasts. The same virus was also isolated from normal feline embryo cells by treatment with iododeoxyuridine. Thus, much to everyone's disappointment, RD114 represents a latent virus of cats which when activated can replicate more readily in human or primate cells.

About the same time as RD114 virus was discovered another group of workers isolated a B-type RNA virus from human milk. This virus was present in the milk of 6 of 10 women from families with a history of breast cancer but only in 7 of 156 women with no history of the disease. This human milk virus has been extensively characterized and in all respects closely resembles mouse mammary tumour (MMT) virus. The RNA of the virus hybridizes extensively with DNA prepared from MMT virus RNA with the aid of reverse transcriptase. Furthermore, this DNA hybridized with 21 of 29 RNA samples from malignant breast tumours but with none of 40 RNA samples from benign human breast tumours or normal breast tissue. These data imply that among the RNA molecules in human breast cancer cells are sequences complementary to MMT virus DNA, the implication being that these RNA molecules appear in the cancer cells because human milk virus is replicating in them. However, these findings by no means prove that human milk virus causes breast cancer.

Even if viruses can be implicated as the causative agents of human cancer, a cure will not be found easily. Many of these viruses could be transmitted vertically thus rendering immunization programmes ineffective. A knowledge of the mechanism of viral replication might permit the isolation of a suitable anti-cancer drug but this is unlikely since this line of approach has not been particularly successful with other viral diseases.

GLOSSARY

Benign An adjective used to describe growths which do not infiltrate into surrounding tissues. Opposite of malignant (q.v.).

Cancer A growth which is not encapsulated and which infiltrates into surrounding tissues, the cells of which it replaces by its own. It is spread by the lymph vessels to other parts of the body. Death is caused by destruction of organs to a degree incompatible with life, extreme debility and anaemia or by haemorrhage.

Carcinoma   A cancer of epithelial tissue.

Fibroblast   A cell derived from connective tissue.

Leukaemia   A blood disease in which there is a great increase in the number of white blood cells.

Lymphoma   A tumour (q.v.) of lymphoid tissue.

Malignant   A term applied to any disease of a progressive and fatal nature. Opposite of benign (q.v.).

Neoplasm   An abnormal new growth, i.e. a cancer.

Neurolymphomatosis   A disease marked by mononuclear cell infiltration of peripheral nerves, especially of legs and wings, of chickens approaching maturity.

Rhabdomyosarcoma   A malignant tumour of striated muscle fibres.

Oncogenic   Tending to cause tumours (q.v.).

Sarcoma   A cancer developing from fibroblasts (q.v.).

Tumour   A swelling, due to abnormal growth of tissue, not resulting from inflammation.

16 The evolution of viruses

Viruses undergo evolutionary change just like any other living organism. Their genomes are subject to mutations at the same rate as all nucleic acids and where conditions enable a mutant to multiply at a rate faster than its fellows, that mutant virus will be selected and will succeed the parental type.

In any discussion of evolution one naturally starts at the earliest time possible and it is pertinent to ask where viruses first came from. The absence of any fossil records of viruses and scarcity of other evidence has not, of course, prevented scientists speculating about the origins of viruses! The two prevailing opinions take into account the parasitic existence of viruses today:

1. viruses have arisen from degenerate cells which have lost the wherewithal for free-living, or
2. viruses have arisen from pieces of cellular nucleic acid which have escaped from the cell. The molecular biology of bacteriophages and their prokaryotic host cells differs considerably from that of viruses of eukaryotes and their host cells, to the extent that it is not possible to grow bacteriophages in eukaryotic cells or eukaryotic viruses in bacteria. Thus it appears that phages and viruses of eukaryotes have arisen independently.

Whatever their origins, viruses have been a great biological success, for no group of organisms has escaped their attentions. In Chapter 11 we discussed the various ways that viruses interact with their hosts and we saw how viruses cause a variety of changes ranging from the imperceptible to death. Evolution of any successful parasite has to ensure that the host also survives. The various virus-host interactions alluded to earlier can be thought of as ways in which this problem is being solved.

In this chapter we shall discuss firstly the evolutionary implications of the distribution of morphologically similar viruses throughout a range of different hosts and secondly, examples of virus evolution which have occurred in relatively recent times.

SPREAD OF MORPHOLOGICALLY SIMILAR VIRUSES

The commonest occurring bacteriophage (see Table 2.2) has double-stranded DNA packaged in a particle of the 'head' and 'tail' type which reaches it zenith of complexity in the T-even phages. This type of phage infects a wide variety of the bacterial species. There are a range of variations in head size and tail length and it is easy to construct a gradient of variation from the simple to the complex after electron microscopic examination of the phages (see Fig. 2.12). This does not necessarily indicate any evolutionary relationship but it says that either these phages arose many times independently or that the basic pattern has been subjected to evolutionary modification. It is interesting to note that the 'head' and 'tail' virus particle has never been found in eukaryotes.

Another example of a widespread group of viruses are the rhabdoviruses of eukaryotes (Fig. 2.10). Their 'bullet-shape' makes them easy to identify. They have a lipid envelope and are in a class V of the Baltimore Scheme. Rhabdoviruses infect both plants and animals, and are found in invertebrates and cold- and warm-blooded vertebrates (but like all viruses each rhabdovirus serotype infects a restricted range of hosts). Since these viruses are morphologically indistinguishable it is tempting to speculate that they have arisen once and subsequently spread in a truly remarkable fashion. The rhabdoviruses also present us with an interesting link between viruses of the animal and plant kingdoms since representative species (e.g. lettuce necrotic yellows virus) multiply in both the insect vector and the plant it feeds on.

EVOLUTION OF MEASLES VIRUS

Measles virus is a paramyxovirus belonging to class V of the Baltimore scheme. It infects only man, and the infection results in life-long immunity from the disease. Studies have been made of the frequency of the disease in island populations (Table 16.1). There is a good correlation between the size of the

Table 16.1 Correlation of the occurrence of measles on islands with the size of the population

Island group	Population $\times 10^{-3}$	New births per year $\times 10^{-3}$	% months with measles 1949-1964
Hawaii	550	16.7	100
Fiji	346	13.4	64
Solomon	110	4.1	32
Tonga	57	2.0	12
Cook	16	0.7	6
Nauru	3.5	0.17	5
Falkland	2.5	0.04	0

population and the number of cases of measles recorded on the island throughout the year. A population of at least 500 000 is required to provide sufficient susceptible individuals (i.e. births) to maintain the virus in the population. Below that level the virus will eventually die out unless it is re-introduced from an outside source.

On the geological time-scale man has evolved recently and has only existed in large populations in comparatively modern time. In answer to the question about where measles virus was in the days of very small population groups, we can certainly conclude that it did not exist in its present form. It may have had another strategy of infection such as to persist in some form and infect the occasional susceptible passer-by, but we have no evidence of this capability. However, F. L. Black has made speculations based upon the antigenic similarity between measles, canine distemper and rinderpest. The latter two viruses infect dogs and cattle respectively which have been commensal with man since his nomadic days. Black suggests that these three viruses have a common ancestor which infected prehistoric dogs or cattle. The ancestral virus evolved to the modern measles virus when changes in the social behaviour of man gave rise to populations large enough to maintain the infection. This evolutionary event would have occurred within the last 6000 years when the river valley civilizations of the Tigris and Euphrates were established.

EVOLUTION OF MYXOMA VIRUS

Myxoma virus belongs to the pox virus family (class I of the Baltimore scheme) and causes a benign infection in its natural host, the South American rabbit, producing wart-like outgrowths (benign tumours) as the only visible evidence of virus multiplication. In the European rabbit myxoma virus causes myxomatosis, a generalized infection with lesions over the head and body surface, which is usually fatal. In nature the disease is spread by virus carried on the mouthparts of the mosquitoes which feed on rabbits, or the rabbit flea. However, this virus does not multiply in the vector.

Myxoma virus was released in England and Australia upon a wholly susceptible host population of the European rabbit in an attempt to eradicate the rabbit as a serious agricultural pest. This experiment in nature was carefully studied with respect to the changes occurring in the virus and the host populations. As we shall see it provides an object lesson in biological control.

In the first attempts to spread the disease in Australia, myxoma-infected rabbits were released in the wild but despite the virulence of the virus and the presence of susceptible hosts the virus died out. It was realized later that this failure was due to the scarcity of mosquito vectors whose incidence is seasonal. When infected animals were released at the peak of the mosquito season an epidemic of myxomatosis followed. Over the next two years the virus spread 3000 miles across Australia and even across the sea to Tasmania. However, during this time it became apparent that fewer rabbits were dying from the disease than at the start of the epidemic. The investigators found two significant facts. Firstly, they compared the virulence of the original virus with virus newly isolated from wild rabbits by inoculating isolates into standard laboratory rabbits. They found that (a) rabbits took longer to die, and (b) a greater number of rabbits recovered from infection. From this it was inferred that the virus had evolved to a more avirulent form (Table 16.2). The explanation was simple: mutation produced virus variants which did not kill the rabbit as quickly as the parental virus. This meant that the rabbits infected with the mutant virus survived to be bitten by the vectors for a longer period than rabbits infected with the original strain. Hence the mutant would be transmitted to a greater number of rabbits. In other words there was a strong selection pressure in favour of mutants which survived in the host in a transmissible form for as long as possible.

The second fact concerned the rabbits themselves and the question was raised that rabbits which were genetically resistant to myxomatosis were being selected. To test this hypothesis a breeding programme was set up in the laboratory. Rabbits were infected and survivors were mated and bred. Offspring were then

Table 16.2 Evolution of avirulence in myxoma virus after its introduction into Australia in 1950

Mean survival time (days)	Mortality rate (%)	Year of isolation			
		1950-1951	1952-1953	1955-1956	1963-1964
<13	>99	100	4	0	0
14-16	95-99		13	3	0
17-28	70-95		74	55	59
29-50	50-70		9	25	31
	<50		0	17	9

infected and the survivors mated and so on. Part of each litter was tested for its ability to resist infection with a standard strain of myxoma virus. The result confirmed that the survivors of each generation progressively increased in resistance.

This work shows how evolutionary pressures set up a balance between a virus parasite and its host which ensures that both continue to flourish. This fact remains a stumbling block to the advocates of biological control of pests which attack man and his animals or plants.

EVOLUTION OF INFLUENZA VIRUS

Background

The three types of influenza virus are distinguished by the antigenicity of their nucleoproteins and are called type A, B, or C. Type A causes the world-wide epidemics (pandemics) of influenza, and both types A and B cause epidemics during the winter. Type C causes minor upper respiratory illness and will not be discussed further. Resistance to infection is determined by whether or not the immune system has been previously exposed to the same virus. The viral antigens relevant to protective immunity are the external haemagglutinin (HA) and neuraminidase (NA) glycoprotein spikes (see Fig. 2.10). Earlier in this chapter we mentioned that infection with measles virus resulted in lifelong immunity to measles. Why then is it common experience for people to suffer several attacks of influenza in their lifetime? The answer is that influenza A and B viruses are continuously evolving new HA's and NA's against which previously acquired immunity is ineffective. How and why this happens is discussed later.

Antigenic drift

Influenza A viruses have been isolated from man since the discovery in 1932 that they could infect ferrets, although embryonated eggs and tissue cultures are used now. Each new isolate was tested serologically with antisera to all other known influenza strains. It soon became apparent that the more recent isolates had slightly different antigens from earlier strains. Next it was realized that the 'old' strains were no longer present and only the new strains could be isolated. This phenomenon is aptly called *antigenic drift* (Fig. 16.1). It rests on the assumption that in nature influenza strains carrying new antigenic determinants arise by

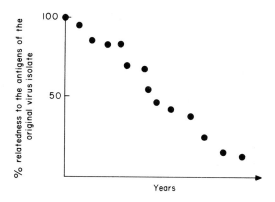

Figure 16.1 Diagram showing antigenic drift. This could represent the HA or NA of either influenza type A or type B strains. Each point represents a strain isolated in a different year.

natural selection of mutants occurring in the current human influenza virus population.

Evolution cannot occur at such a fast rate without efficient selection for a favourable mutant. How this might come about was shown by experiments in which virus was grown in cell culture in the presence of critical amounts of antibody (Fig. 16.2). After 7 passages into new cultures containing antibody, the resulting

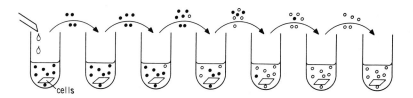

Figure 16.2 Antigenic drift in a test tube. Cultures containing antibody insufficient to completely neutralize the virus are inoculated. Progeny virus is transferred to new cultures which also contain antibody. Mutants (o), resistant in some degree to the antibody, arise spontaneously and have a selective advantage. They gradually become dominant in the population.

virus was used to immunize rabbits and compared in reciprocal haemagglutination inhibition tests with the original strain. The HAs were now very different. This 'evolution in a test-tube' is thought to mimic what occurs in nature in our respiratory tracts.

However, the antigenicity of another virus which initially causes a respiratory infection, measles virus, is unchanging. Why then does measles virus not evolve like influenza virus? The essential difference is that only measles develops into a generalized infection with virus in the blood. This viraemia is open to the full force of the immune system and the resulting immunity can effectively get rid of both parental virus and any variants that arise. Influenza virus remains in the respiratory tract which cannot mount such a rigorous immune response as the circulatory system. This apparent defect is a physiological compromise to prevent over-reaction by the immunologically active cells to the vast amounts of foreign antigens which we continuously inhale.

Antigenic shift

It became clear from examining strains isolated over a period of 40 years that the evolution of influenza type A virus was not solely a continuous process of antigenic drift but rather was broken up by the mysterious appearance of new strains, called subtypes, with an HA or NA or both totally unrelated to those of the previous years (Fig. 16.3). By contrast, influenza type B undergoes antigenic drift only.

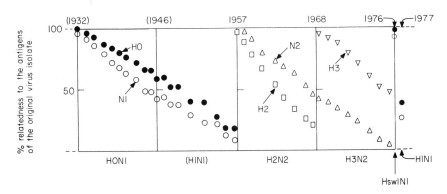

Figure 16.3 Diagram showing antigenic shift. In each of the years written in full, a haemagglutinin (•) or neuraminidase (o) appeared, which was antigenically unrelated to the strains of the preceding year. The haemagglutinin of the first human subtype isolated in 1932 is designated H0. Until recently it was thought that H0 and H1 (in parentheses above) were distinct serotypes but the current view is that these have arisen by antigenic drift from the Hsw1 of the strain causing the pandemic of 1918. The subtypes H2 and H3 are the result of antigenic shift, while the 1976 and 1977 strains may have resulted from recurrence of previous strains and this is discussed in the text.

Table 16.3 Correlation between the occurrence of major outbreaks of influenza with the appearance of new subtypes

Year of epidemic or pandemic	Designation subtype
1889	H2N2*
1900	H3N2*
1918 ("Spanish 'flu")	Hsw1N1*
1929	H0N1*
1946	H1N1†
1957 ("Asian 'flu")	H2N2†
1968 ("Hong Kong 'flu")	H3N2†
1977 ("Russian 'flu")	H1N1†

*Serological evidence
†Virus isolations

In addition to the subtypes discovered since human influenza viruses were first isolated there is serological evidence (explained in the section on 'cycling' below)

214

that man has in the past been infected by subtypes related to modern H2N2 and H3N2 viruses of man and Hsw1N1 virus of pigs.

The significance of these new subtypes is apparent in the close correlation of their appearance with the occurrence of major outbreaks of influenza (Table 16.3). This is not too surprising since immunity developed to the previous subtype will provide no protection against a virus carrying new antigen(s). Various explanations have been advanced to explain antigenic shift and we shall consider the following.

Critical alterations of antigenic structure

This postulates that a mutation results in the substitution of an amino acid at a critical part of an antigenic site which causes the protein to adopt an entirely new configuration and hence antigenicity. The theory is an extreme version of antigenic drift. A prediction which follows from this hypothesis is that the bulk of the protein should have a sequence identical with the strain from which it evolved. This is not borne out by biochemical analysis as 'maps' of peptides produced by digestion with trypsin show considerable differences between, for example, 1967 H2 and 1968 H3.

Explanations which involve other hosts

Some of these hypotheses depend upon the ability of human influenza strains to infect animals and more importantly on the fact that many different subtypes of influenza A viruses have birds, pigs and horses as their natural hosts. Influenza type B infects only man. There is thus a correlation between the non-existence of type B strains which infect animals and the non-occurrence of antigenic shift or pandemics caused by this virus.

Change of host. The simplest explanation of antigenic shift is that a non-human strain acquires the ability to infect man. This would account readily for the isolation in 1957 of a virus which had an HA and NA totally different from the strain around in the previous year. (A shift involving a single antigen implies a familial relationship with a strain in the preceding year.) In 1976, the same swine influenza virus was isolated from both pigs and pig-farmers in the Eastern U.S.A., demonstrating that exchange of influenza viruses between species does take place. Surprisingly no epidemic resulted even though the population had no immunity to the virus (Hsw1N1). It is surmised that the virus lacked the ability of being transmitted, a hitherto undocumented property.

Cycling. This theory is based upon the presence of influenza antibodies in sera obtained from people who were alive long before 1932, when techniques for isolating viruses became available. Sera taken before 1957 when the H2N2 subtype first appeared were kept frozen and then tested for antibodies to the modern H2N2 and H3N2 viruses. People who were alive in 1889, but not 1888, had H2N2 antibodies, suggesting that they had been infected with an H2N2 virus in 1889. Similarly it was inferred that an H3N2 virus was around in 1900.

The theory suggests that strains 'go into hiding', perhaps in another host, where they remain until people with immunity have died and there is a substantial population of susceptibles. The virus then emerges and can commence infection in man. The cycle for H2N2 and H3N2 subtypes turned completely in 60-70 years which is about the average life expectancy (Fig. 16.4). However the

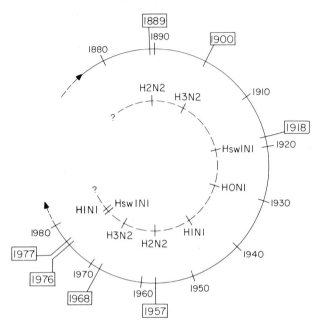

Figure 16.4 Summary of the evidence which suggests cycling of human influenza type A viruses. The time-scale in the outer circle is marked with the year that a new subtype emerged: 1918.

appearance in 1977 of an H1N1 strain identical by serology and molecular hybridization with the 1950 H1N1 suggests that sufficient susceptible people may accumulate in only 27 years.

Genetic assortment. The genome of influenza A (and B) viruses consists of 8 single-stranded RNA segments each of which is the complement of a mono-cistronic message. When a cell is infected simultaneously with more than one strain, newly synthesized RNA segments assort virtually at random to the progeny (Fig. 16.5). Such hybrid strains are genetically stable. They have been produced experimentally in cell culture and in whole animals. It was shown that hybrid strains also had the ability to spread naturally from infected to susceptible animals, an essential property in the evolution of a new strain. Viruses which could have arisen by genetic assortment between human and animal strains (e.g. having the HA of an animal strain and the NA of a human strain) have been isolated in nature. However *proof* that they did arise by genetic assortment is lacking and may always be unavailable.

216

Summary

Type A influenza viruses evolve rapidly and in all probability by two different processes. There are several ways in which antigenic shift could occur but there is only circumstantial evidence available. The mechanisms discussed are not necessarily exclusive and all may operate on different occasions.

Type B influenza viruses do not undergo antigenic shift although they can undergo genetic assortment. The absence of non-human type B strains may be the essential ingredient missing. Type A and B strains do not form hybrids.

Other viruses are subject to antigenic drift in nature, notably the virus of foot-and-mouth disease. In the laboratory, viruses of stable antigenicity give rise to antigenic variants under the selective pressure of antibody. Clearly the capacity for antigenic evolution is there even if conditions do not normally permit it.

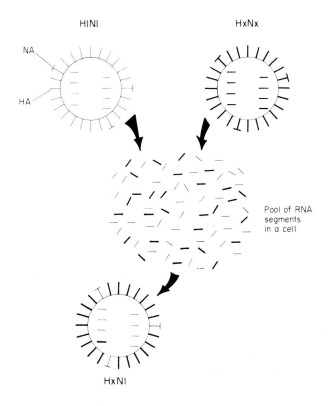

Figure 16.5 Explanation of antigenic shift in influenza type A viruses resulting from simultaneous infection of a cell by the parental viruses. The 8 genome segments from each parent assort independently to progeny virions. In the example shown, only the segment carrying the genetic information for the HA assorted to the progeny. A total of 256 genetically different types of progeny virus are possible.

Postscript on the World Health Organization

The W.H.O. is the body which co-ordinates viral epidemiology world-wide. There is a network of National and International Reference Centres which study a variety of viruses and in particular are monitoring the appearance of new, potentially pandemic, strains of influenza. Valuable information is gained on virus evolution and vaccine manufacturers can be alerted to a new strain in time to protect vital sections of the community in the face of a pandemic.

Another aspect to virus evolution is the near eradication of the virus causing smallpox in man which has also been carried out under the auspices of the W.H.O. When successful, this will be the first instance of the deliberate extinction of a human pathogen.

17 Trends in virology

Bacteriophages first attracted the attention of Max Delbruck in the 1930's. A theoretical physicist by training he became increasingly interested in the nature of the gene and thought that a study of bacterial viruses might provide the vital clues. The fact that Delbruck himself, as well as numerous others who followed in his footsteps, received for his phage work the ultimate accolade, a Nobel prize, attests to his foresight. The golden years for phage research were undoubtedly the 1950's and 1960's. In the first of these decades the excitement centred on the genetics of bacteriophages and in the second attention was focused on viral nucleic acid replication and its regulation. In the last decade the interest in bacteriophages has largely died out. Why should this be?

There are two major reasons for the decline in phage research. Firstly, bacteriophages are of little concern economically except where there is large-scale cultivation of bacteria. In the days of the acetone-butanol fermentation, when wartime economies demanded that energy be saved by only partially sterilizing the growth medium, culture lysis was a major problem. However, modern petrochemical technology has replaced the acetone-butanol fermentation, and others like it, and the only large-scale microbiological processes with significant phage problems are antibiotic manufacture and cheesemaking. Actinomycetes are widely used for the production of antibiotics and actinophages are reputed to cause considerable problems. However, the extent of the problem is hard to gauge as the pharmaceutical trade is particularly reticent about admitting its problems. By contrast, the dairy industry freely admits to the problems posed by phage, particularly the lysis of cheese starters. These problems stem from a number of sources, e.g. the widespread polylysogeny in starter strains, the high calcium content of milk which favours phage adsorption, and the use of open vats. Clearly these problems posed by phages are not sufficient to maintain the huge momentum of phage research which built up in the previous two decades.

A more important reason for the decline in phage research was the growing realization in the 1960's that animal viruses could be manipulated just as easily as bacteriophages. The biochemical and genetical techniques first developed by phage workers were rapidly adapted by animal virologists such that many animal viruses are now amenable to experimentation. Curiously, whereas the growth of animal virology can be traced back to the development of a plaque assay for animal viruses, the development of the local lesion assay did not promote similar advances in plant virology. The explanation is simple. The plaque assay method for titrating animal viruses depends upon the ability of viruses to infect monolayers of cells in tissue culture. In effect this permitted the study of animal viruses outside the whole animal. The local lesion assay, on the other hand, required the whole plant and attempts to grow plant cell monolayers have not been particularly successful. Consequently the study of plant viruses has lagged behind that of bacterial and animal viruses.

What will be the trends in virology in the near future? Clearly the prevention of virus diseases in man, his domestic animals and his crops still remain the major aims of virologists. As far as animal diseases are concerned, some will be eliminated by effective vaccination programmes, as smallpox has been, but the prospects with others, such as influenza and the common cold, are less promising (see page 184). One possible chemotherapeutic agent is human interferon (see page 189). Clinical trials have shown the potential of human interferon in stemming viral disease but such studies have been limited because of the vast quantity of interferon needed and the limited stocks available. However, several groups around the world are exploiting recombinant DNA technology (see later) in an effort to clone the human interferon gene in *E. coli*. If successful, this could pave the way for the large-scale production of interferon for clinical use. Plant viruses are no less important for there are serious virus diseases of virtually all crops whether grown for a fruit, a vegetable, a grain, a beverage or a fibre. Such diseases often do not kill the affected host but they can reduce its vigour leading to a greatly decreased yield. Unfortunately, prospects for chemotherapy of virus-infected plants are not bright. Although there is some evidence for an interferon-like substance in plants its study lags far behind that of the animal interferon. Even if such a substance could be purified from plants it is not easy to visualize how it could be effectively used.

VIRUSES AND GENE MANIPULATION

The purification of restriction endonucleases and characterization of their mode of action (page 95) rapidly led to the development of sophisticated techniques for gene manipulation. The key feature of those techniques is the ability of certain of these restriction endonucleases to produce staggered cuts at certain well-defined sites on DNA molecules (see Fig. 6.18 for examples). The way in which these enzymes can be used is shown in Fig. 17.1. DNA's from two different sources

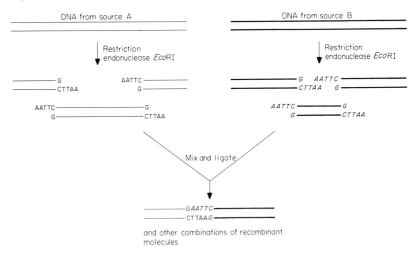

Figure 17.1 *In vitro* recombination of two DNA fragments.

are cleaved with the same endonuclease, e.g. *Eco* RI, to produce fragments with complementary single-stranded tails. By incubating mixtures of the two fragments under annealing conditions and ligating with DNA ligase it is possible to produce mixed dimers, trimers etc.

Although it is relatively easy to introduce fragments of DNA into bacteria by the process of transformation, such fragments will fail to replicate and will be lost from the cell. To circumvent this problem the DNA fragment is inserted into a *vector* (or *cloning vehicle*) which is simply a DNA molecule capable of autonomous replication. Commonly used vectors are bacterial plasmids and bacteriophage λ. Bacteriophage λ has two advantages as a vector. Firstly, the hybrid DNA can be prepared in great quantities since each bacterial cell can produce several hundred λ DNA copies and the hybrid DNA can be purified easily as a component of bacteriophage particles. By using λ mutants that make the host lysis defective it is possible to increase the phage yield up to ten times and in addition the phage remain in the bacterial cell until artificially lysed. Secondly, the detailed knowledge of the λ genome is particularly useful for it is possible to build mutations which increase the transcription of the foreign DNA insert.

The basic disadvantage of bacteriophage λ as a cloning vehicle is that it has five target sites for *Eco* RI (Fig. 17.2). Regions B and C of the genome are not necessary for *lytic* growth of λ and so can be deleted. This was done by cleaving

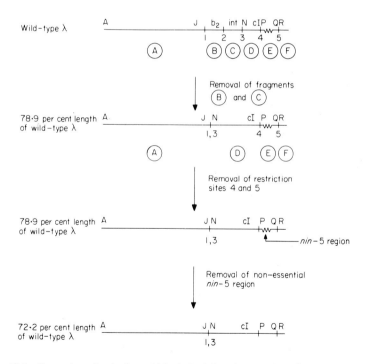

Figure 17.2   Generation of a cloning vehicle derived from bacteriophage λ. The vertical bars represent the five sites for *Eco* RI cleavage and A-F the fragments which would be produced. A, J, b2, int, N, cI, P, Q and R are viral genes.

λ DNA with *Eco* RI, separating the fragments by electrophoresis in agarose gels, and joining fragment A to fragments D, E and F. This gives a λ molecule with 21.1% less DNA and only 3 *Eco* RI substrate sites. Two of these 3 sites (nos. 4 and 5) were removed by mutation yielding a DNA molecule with only one *Eco* RI site. The shortest λ-DNA molecules which still produce plaques on transfection are nearly 25% deleted. However, if too much DNA, even non-essential DNA, is deleted from the λ genome it cannot be packaged into virions. Deletion of the non-essential *nin*-5 region (6.1% of the λ genome) as well as the B and C fragments yields a DNA molecule which cannot give rise to plaque-forming particles. Such a molecule is ideal for cloning DNA. By cleaving the deleted λDNA, mixing with foreign DNA similarly cleaved, annealing and transfecting we can select plaque-forming λ particles and these must contain inserts.

Cloning genes in animal cells

Methods analogous to those employed for cloning DNA in *E. coli* have been developed to propagate foreign DNA in eukaryotic cells. These involve simian virus 40 (SV40) for cloning in animal cells and cauliflower mosaic virus (CaMV) for cloning in plants. Because the initiation site of SV40 is the only *cis*- required function for DNA replication, a segment of viral DNA containing the origin of replication can replicate in the presence of a helper SV40 DNA molecule which provides the necessary *trans*- functions. Any foreign DNA that is covalently joined to the viral segment containing the origin should also be replicated in susceptible (monkey) cells. In addition, the ability of SV40 to integrate into the chromosomal DNA may allow expression of the genes it contains.

An SV40-bacterial DNA hybrid has been constructed in which a 1300 base-pair DNA segment in the late region of the SV40 genome has been replaced by an 870 base-pair bacterial DNA segment containing the suppressor tRNA gene su+ III. After co-infection of monkey kidney cells at the non-permissive temperature with the SV40-su + III molecules and a temperature-sensitive mutant of SV40, to supply late gene functions necessary for virus production, the recombinant DNA was replicated and encapsulated into the virions. Structural studies on the hybrid showed that the *E. coli* DNA insert suffered no observable sequence alterations. In addition, monkey cells infected with SV40-su + III virus transcribed the bacterial DNA fragment but did not produce functional suppressor tRNA. It has also been possible to obtain SV40-transformed rat cell lines carrying the bacterial DNA linked to the SV40 vector.[†]

Cloning genes in plant cells

From the above it is clear that SV40 has great potential as a cloning vehicle for animal cells although the techniques are not nearly as well developed as for bacteriophage λ. The potential of CaMV, for cloning in plant cells, remains to be seen. The genome of CaMV consists primarily of a double-stranded circular

---

[†] Note added in proof: Recently a number of groups have observed expression of foreign genes, e.g. mouse β-globin, cloned in SV40 DNA.

molecule with a mol. wt. of 4.6×10^6. However, the DNA contains at least 3 single-stranded regions and may even contain stretches of RNA-DNA hetero-duplex. CaMV has a very limited host range and with few exceptions infects only members of the Cruciferae. Infections are systemic and spread to a majority of plant cells in a few weeks. Both the virions and naked viral DNA are infectious and the ability to infect with viral DNA will be of great advantage in developing a cloning system. Recombinant DNA molecules have been constructed from CaMV DNA and bacterial plasmids but after replication in *E. coli* the viral DNA does not replicate in plant cells. Clearly, much remains to be done to develop CaMV as a cloning vehicle.

ELIMINATION AND CURE OF VIRUS DISEASES

Elimination of virus diseases

Smallpox has almost certainly been totally eliminated: the last naturally occurring case was recorded in Ethiopia in October 1977. Yet less than 200 years ago mortality from smallpox in England reached 25% of all children born, and in India in 1950 over 41 000 people died. The decline of smallpox in those countries where it was recently endemic has been a triumph for the vaccination programme administered by the World Health Organisation. Today they offer a reward of $1000 to anyone who finds a 'natural' case of smallpox. Smallpox is the only disease which has ever been deliberately eliminated and this experience has served to emphasize that successful elimination is possible only if a disease fulfils certain criteria. The causal virus must not persist in the body after the initial infection, there must be no animal reservoir from which re-infection of man can occur, vaccination must provide effective and long-lasting immunity, and the viral antigens must not change. Apparently smallpox fulfils all of these criteria, although the isolation from monkeys of whitepox virus, which is antigenically similar to smallpox, requires investigation to ensure that it will not replace smallpox as a pathogen of man.

A problem arises with laboratory stocks of smallpox virus. As vaccination is no longer required by most countries, individual and herd immunity falls so that the population becomes increasingly susceptible to the danger of an escaped laboratory virus. We are faced with the dilemma of deliberately destroying a virus and losing, for ever, 75×10^6 mol. wt. units of irreplaceable genetic information. In agricultural circles it is fully realized that the primitive or 'unimproved' stocks carry genes of value (e.g. resistance to disease) which have been lost in the selection for other attributes, and that the stocks should therefore be preserved in case the lost characters need to be replaced. We have no way of knowing the value of the smallpox virus genome but it codes for many enzymes which duplicate those found in eukaryotic cells and which conceivably in the future, could be used to restore deficient cells to normal by genetic engineering.

In October 1978, the Department of Health, Education and Welfare announced their intention of eliminating measles virus from the U.S.A. Death results from respiratory and neurological causes in about 1 of every 1000 cases and encephalitis also occurs in 1 of every 1000 cases, survivors of the latter often having permanent brain damage. Are the criteria for successful elimination met by measles virus? It multiplies only in man; there is a good live vaccine (95% effective) and only one serological type of virus is known. Usually measles virus causes an acute infection but, rarely (1 of every 10^6 cases), the virus persists and reappears some 2-6 years later as a progressive disease of the brain called subacute sclerosing panencephalitis (SSPE). However measles virus can only be recovered with difficulty from infected tissue and SSPE is in that sense a non-transmissable disease.

To eliminate measles it is necessary to achieve a high immunization level, especially in children, and the present programme is concentrating on achieving this. As part of this aim, some schools require that children are immunized as a condition of their attendance and the ethical question is raised of the freedom to refuse medical treatment without needing to plead special medical or religious reasons.

Another candidate for elimination could well be rubella virus since it causes congenital abnormalities when women are infected in the first three months of pregnancy. However all immunization programmes are expensive and can only be implemented if the country can afford to foot the bill.

Cure of virus diseases

It is always salutary to a virologist to be reminded that there is little that can be done to alter the course of a virus infection. Today we have a handful of highly toxic drugs which can only be used during life-threatening infections caused by DNA viruses or in some superficial infections of the skin or eye. The problem resides in the intimate life-style of the virus with its host cell and in the difficulty of pinpointing those events which are unique to the virus and yet amenable to effective anti-viral compounds which are not toxic to the individual. Understanding of those initial events of the multiplication cycle, attachment and penetration, would be a worthwhile goal since a virus which is excluded from a cell cannot multiply. Use of the universal anti-viral compound, interferon, is restricted only by the problem of producing enough of it (chapters 12 & 14) but theoretically the problem can be solved by cloning the interferon gene. We wait anxiously to see if this can be achieved in practice.

Another advance which may prove useful in treatment of acute virus diseases is monoclonal antibody. A plasma cell which is synthesizing high affinity neutralizing antibody is made immortal by fusion with a cell from a continuous line. One has only to collect the culture fluids from the resultant monolayer of cells to have unlimited amounts of antibody with a specific activity higher than an animal could ever produce. Passive immunization with such antibodies may be necessary to meet the threat of highly lethal viruses such as Lassa fever, Ebola and Marburg (see page 227), particularly in view of the obvious difficulties in handling them for the preparation of vaccines.

VIRULENCE—THE MAJOR UNSOLVED PROBLEM

There is greater ignorance of virulence than of any other area in the virology of eukaryotes. While we can describe what happens during a virus disease, we have no notion of the molecular events which distinguish otherwise identical virulent and avirulent strains. There can be no doubt that when their nucleic acids are compared by oligonucleotide mapping or even sequence analysis differences will be found, but the essential problem still remains to correlate these (or other properties of the virus) with whatever controls virulence and then to understand in molecular terms how this virus strain interacts with its host to produce a violent end-result while the avirulent strain does not. The analysis is complicated since virulence is the end-product of the very complex reactions of the animal or plant host to the infectious agent and this, by its essence, invalidates experiments done in cultured cells. Our ignorance is so basic that we even lack a concept within which to frame experiments.

ECOLOGY OF VIRUSES

Where a virus causes a serious disease of man, his domestic animals or crops, considerable effort has been devoted towards understanding the ecology of that virus. Obviously in the absence of suitable anti-viral agents to minimize the effects of infection, the only way to control virus diseases is to identify, and where possible to eliminate, vectors and reservoirs of the disease. Thus virus yellows in British sugar-beet crops can be largely controlled by removing weeds, mangold clamps and old beet crops which serve as overwintering reservoirs of both the virus and its aphid vector. In the past relatively little attention has been paid to the ecology of viruses not directly affecting man but this situation is clearly changing for reasons best exemplified by the baculoviruses.

Insect viruses

More than 600 viruses have been isolated from dead or moribund insects although very few have been studied in detail. Since only a small proportion—probably less than 1%—of the total number of insect species have been examined many more viruses of insects remain to be discovered. So far eight groups of insect viruses have been described (Table 17.1). Only the baculoviruses appear to be restricted to insects and because of their specificity they have been considered as biological control agents. Most of the insect pests of world-wide importance in agriculture and forestry are members of the Lepidoptera and Hymenoptera and it is fortunate that baculoviruses are commonly found in species in these groups. Such viruses can cause epizootics in their host population to such a level that very effective natural control is exerted. For example, in recent years large areas of spruce in Wales were attacked by the spruce sawfly. Previously, the sawfly had only been a local pest and the reasons for the widespread population growth are not clear. However, in 1973 a baculovirus was observed to be causing considerable mortality of the sawfly and by 1976 effective natural control had been established.

It is significant that when baculovirus preparations have been applied to insect populations in the field, the degrees of mortality, and therefore control, have been extremely variable. Unfortunately, the reasons for success or failure in such applications are not fully understood and are obviously a hindrance to

Table 17.1 Viruses infecting insects

| VIRUS FAMILY | VIRUS | VIRUS SHAPE | NUCLEIC ACID | SIMILAR VIRUSES FOUND IN | |
				VERTEBRATES	PLANTS
Baculoviridae	Baculovirus	Rod	DNA	NO	NO
Reoviridae	Cytoplasmic polyhedrosis virus	Spherical	RNA	YES	YES
Poxviridae	Entomopox virus	Ovoid	DNA	YES	NO
Iridoviridae	Iridovirus	Spherical	DNA	YES	NO
Parvoviridae	Densovirus	Spherical	DNA	YES	NO
Picornoviridae	Insect enteroviruses	Spherical	RNA	YES	YES*
Rhabdoviridae	Sigmavirus	Bacilliform	RNA	YES	YES
Non-occluded RNA viruses		Spherical	RNA	Possibly	YES

*Turnip yellow mosaic virus, a tymovirus, resembles the picornaviruses.

further development. Under natural conditions in the field, insects are usually exposed to very small doses of virus. The development of a natural epizootic could arise from a natural increase in the amount of virus inoculum or because the defensive mechanisms of the host are inhibited in some way not previously known. Thus the successful creation of artificial epizootics by deliberate release of the virus would depend on the dosage available, its retention of infectivity, and the state of the insect host. It is in this context that the deliberate release into the environment of insect pathogenic viruses should be considered and the resulting ecological danger assessed. The hazard that could arise from large-scale spray application of insect pathogenic viruses can be summed up as the potential for accidental infection of invertebrate or vertebrate animals with the possibility of overt disease or, more insidiously, widespread inapparent infections.

Viruses in water

There are other reasons why the ecology of viruses will assume growing importance. For example, there has been a dramatic world-wide tendency toward urbanization with a concomitant increase in discharge of raw and processed sewage into waterways. These waterways in turn are being increasingly used both recreationally and as sources of potable water. Unfortunately the basic design of sewage treatment plants was elaborated before the nature of viruses was fully appreciated and one consequence is that effluent from sewage plants contains most of the viruses which entered with the raw sewage. Those viruses which are removed frequently end up in sludge which is used for landfill or as an agricultural fertilizer. All too frequently sludge is sprayed on land used for grazing sheep and cattle!

Because of concern about the possible presence of serious viral pathogens in potable and recreational waters, a number of research groups around the world are concentrating on elaborating methods for concentrating viruses from large volumes of water. A number of such devices have been described, one of which can successfully concentrate small numbers of poliovirus particles from 380 litres of water. Undoubtedly the next decade will see a considerable improvement in the design for such concentrators. Unfortunately, methods for the satisfactory recovery of viruses from solids have yet to be developed and this will hinder much ecological work.

Exotic virus infections of man

In recent years the general public has learned to fear the name 'Lassa fever' which they recognize as a highly lethal infectious disease acquired in tropical Africa. There are a number of other notable exotic viruses ('exotic' meaning introduced from abroad) which at this moment pose a potential rather than an actual threat to the general population. All of these viruses are zoonotic, that is they are spread to man from their natural hosts, infected mammals, birds or arthropods (Table 17.2). Man is an accidental host and as sometimes happens in these circumstances the resulting disease is more severe than in the natural infection. Lassa fever is a life-threatening infection with a 35-50% lethality rate,

Table 17.2 Some exotic viruses

Virus	Group	Comment
Smallpox	Poxviridae	Eradicated. Possible risk from laboratory stock
Yellow fever	Flavivirus	Endemic in Central Africa. Capable of invading new areas in the wake of its mosquito vector *Aëdes aegypti*. Excellent vaccine available
Venezuelan equine encephalitis	Alphavirus	Epidemic. Serious morbidity and mortality in children
Chikungunya O'Nyong Nyong	Alphavirus	
Dengue, St. Louis, Murray Valley, Japanese, West Nile and various tick-borne encephalitides	Flavivirus	Usually endemic locally and flares into epidemics
Rift Valley fever	Bunyaviridae	Infects man and cattle in explosive epidemic proportions, spread by arthropods. High mortality and morbidity rates
Rabies	Rhabdoviridae	Mostly endemic although exotic in U.K. Difficult to eradicate when introduced. Spreading in Europe. There is now an effective tissue culture-grown vaccine which can be used as a preventative as well as a therapeutic measure. Few deaths (around 5/year) in Europe
Lassa fever	Arenaviridae	Endemic in certain rodents with direct infection to man via contaminated food or water. Danger from international travel and indistinguishable in its early stages from many other fevers. Only immune serum available as treatment
Marberg and Ebola	? Rhabdoviridae	Some similarities with rhabdoviruses but occur up to 4000 nm in length. Resistance to inhibitors of DNA synthesis indicates an RNA genome. Discovered as a result of infection from preparation of primary monkey cell cultures. Epidemic in 1976 in the Sudan involving 600 people with 50-90% mortality. Could re-emerge. Natural history unknown

whereas its natural host *Mastomys natalensis,* a peridomestic rodent, appears to suffer no ill effects. Because exotic viruses are zoonotic there is no hope for their elimination and only control measures, affecting for example their rodent and arthropod hosts, are contemplated. We understand little about many of these viruses and there is need of investigation into the natural history as well as the properties of the viruses themselves. However, because of the extreme biohazard, they can only be handled in the highest containment laboratories and work progresses slowly as a consequence. The entire story of exotic viruses can be summarized as the result of man's intrusion into environments in which natural selection has not fitted him to live.

THE ORIGIN OF VIRUSES

There are two theories concerning the origin of viruses: they are either degenerate bacteria or vagrant genes. Just as fleas are descended from flies by loss of wings, viruses may be derived from micro-organisms which have dispensed with many of their cellular functions. Alternatively, some nucleic acid might have been transferred accidentally to a cell of a different species and instead of being degraded, as would normally be the case, might have survived and replicated. Although over 40 years have elapsed since these two theories were first proposed we still do not have any firm indications about the origin of viruses. However, techniques for the rapid sequencing of nucleic acids have now been developed and this could provide the necessary breakthrough. Already two phages (MS2 and ϕX174) and one viroid (PSTV) have been sequenced, and in the next decade many more viral genomes will be, and, no doubt, techniques will be improved to permit the sequencing of entire cellular genomes. Once this has been done, computer analysis of the data could provide some interesting clues to the origin of viruses. However, such analyses might well only identify the progenitors of a virus but not indicate whether it arose by degeneracy or gene escape.

Studies on PSTV suggest that it may in fact be a vagrant gene. PSTV has been shown to hybridize with cellular DNA of several uninfected host species (family Solanaceae), and infection of such plants with PSTV does not alter the hybridization pattern. Detailed analysis of the kinetics of hybridization indicates that PSTV hybridizes to unique DNA sequences. At least 60% of PSTV is represented by sequences in the DNA of solanaceous host plants but, more interesting, the DNA from plants not known to be hosts of PSTV also contain sequences related to part of PSTV. Phylogenetically, the more distant a plant species from the Solanaceae the fewer PSTV-like genes its DNA will contain. Although by no means proof, these results do suggest that viruses are derived from genes in the solanaceous hosts. One possibility is that they are derived from the DNA inserts which have recently been found in the middle of functional genes of some eukaryotes.

It is unlikely that all the viruses currently known have evolved from a single progenitor. Rather, viruses have probably arisen numerous times in the past by one or both of the mechanisms outlined above. However, once formed viruses would be subject to evolutionary pressures just like prokaryotic and eukaryotic organisms. One process which must contribute significantly to virus evolution is recombination between two unrelated viruses. Thus the *Salmonella* phage P22 can recombine with the morphologically unrelated *Salmonella* phages *Fels* 1 and *Fels* 2, as well as coliphage λ, to yield novel hybrid phages. Such hybrids are easily detected by growing P22 on Fels 1 lysogens, for example, and plating the progeny virus on Fels 1 lysogens resistant to P22. The only plaque-forming particles observed are hybrid phages with the appearance of Fels 1 but carrying a considerable portion of the genome of P22 including the immunity region. Such illegitimate recombination events have so far only been detected in bacteriophages but undoubtedly also occur with viruses of animals. Since recombination between RNA molecules has only been detected between poliovirus genomes and between foot-and-mouth disease virus genomes, it must be assumed for the time being that RNA viruses are not subject to such sudden, gross changes. However, the potential for genetic changes between related viruses with segmented genomes is enormous.

18 The classification and nomenclature of viruses

Listed below are the major groups of viruses together with a brief description and sketches of the particles (*not* to scale). Viral taxonomy is still in progress and many viruses are as yet unclassified. The list is subdivided for convenience into viruses which infect vertebrates and other hosts, vertebrates only, invertebrates, plants and bacteria, although certain viruses cross such boundaries.

It can be seen that the list does not reflect the distribution of viruses amongst their various hosts but the emphasis on research into viruses of medical or veterinary importance, plus those for which we have the good fortune to have excellent animal cell culture systems. However, work is expanding into the other areas which will redress the situation and make this list in need of continuous revision.

VIRUSES MULTIPLYING IN VERTEBRATES AND OTHER HOSTS

Note that some or all of the Reoviridae, Bunyaviridae, Rhabdoviridae and Togaviridae multiply in *both* vertebrates and other hosts. Other families include genera which multiply solely in vertebrates.

Family: Iridoviridae (Class I).

 Double-stranded DNA of mol. wt. $130\text{-}160 \times 10^6$. Particle is a 130-300 nm icosahedron with several shells. Contains enzymes. Some contain lipid. Cytoplasmic multiplication.

Genus: Iridovirus (iridescent viruses of insects).
Other probable genera: Cytoplasmic icosahedral DNA viruses of vertebrates (e.g. African swine fever virus, lymphocytis virus of fish, frog virus 3) may belong to the same family but not to the genus Iridovirus.

Family: Poxviridae (Class I).

 Double-stranded DNA of mol. wt. $130\text{-}240 \times 10^6$. Largest viruses 170-260 × 300-450 nm. Complex structure composed of several layers. Contains lipids and enzymes including an RNA polymerase. Cytoplasmic multiplication.

230

Genera: Orthopoxvirus (vaccinia and related viruses)
        Avipoxvirus (fowlpox and related viruses)
        Capripoxvirus (sheep pox and related viruses)
        Leporipoxvirus (myxoma and related viruses)
        Parapoxvirus (milker's node and related viruses)
        Entomopoxvirus—poxviruses of insects.

} poxviruses of vertebrates.

Family: Parvoviridae (Class II).

Single-stranded DNA of mol. wt. 1.5-2.2 × 10^6. Particle is a 18-26 nm icosahedron. Multiplication has a nuclear stage.

Genera: Parvovirus (latent rat virus group)—viruses of vertebrates.
        Densovirus—viruses of insects.
        Adeno-associated virus group—viruses of vertebrates. Particles contain complementary DNA molecules which form a double strand upon extraction. Require helper adenovirus.

Family: Reoviridae (Class III).

10-12 segments of double-stranded RNA of mol. wt. 10-16 × 10^6. Particle is a 60-80 nm icosahedron. Has an isometric nucleocapsid with transcriptase activity. Cytoplasmic multiplication.

Genera: Reovirus—viruses of vertebrates.
        Orbivirus—viruses of vertebrates, but multiply in insects also.
Probable genera: Rotaviruses of vertebrates. Cytoplasmic polyhedrosis viruses of insects. Clover wound tumour and Fiji disease viruses of plants (probably two distinct genera).

Family: Picornaviridae (Class IV).

Single-stranded RNA of mol. wt. 2.5 × 10^6. Particle is 20-30 nm with cubic symmetry. Multiplication is cytoplasmic.

Genera: Enterovirus (acid-resistant, mainly viruses of enteric tract).
        Rhinovirus (acid-labile, mainly viruses of upper respiratory tract but also foot-and-mouth disease virus).
Possible genus: Calicivirus (vesicular exanthema of swine and related viruses).

Family: Togaviridae (Class IV).

Single-stranded RNA of mol. wt. 4 × 10^6. Enveloped particles 40-70 nm diameter contain an icosahedral nucleocapsid. Haemagglutinate. Cytoplasmic.

Genera: Alphavirus (group A arboviruses e.g. Semliki Forest virus).
Flavivirus (group B arboviruses e.g. yellow fever virus).
Pestivirus (hog cholera and related viruses).
Rubivirus (rubella virus).

Family: Rhabdoviridae (Class V).

 Single-stranded RNA of mol. wt. 3.5-4.6 × 10⁶ complementary to mRNA. The bullet-shaped or bacilliform (130-300 × 70 nm) particle is enveloped with 10 nm spikes. Inside is a a helical nucleocapsid with transcriptase activity. Cytoplasmic.

Genera: Vesiculovirus (vesicular stomatitis virus group) viruses of vertebrates and insects.
Lyssavirus (rabies virus group) viruses of vertebrates.
Possible genera:
Exemplified by: Haemorrhagic septicaemia of trout virus (viruses of vertebrates). Sigmavirus group (viruses of insects). Lettuce necrotic yellows, potato yellow dwarf and many other viruses of plants.

Family: Bunyaviridae (Class V).

 3 segments of single-stranded RNA of mol. wt. 6 × 10⁶. Enveloped 100 nm particles with spikes and an internal ribonucleoprotein filament 2 nm wide. Cytoplasmic.

Genus: Bunyavirus (Bunyamwera and serologically closely related viruses).

VIRUSES MULTIPLYING ONLY IN VERTEBRATES

Family: Herpetoviridae (Class I).

 Double-stranded DNA of mol. wt. 54-92 × 10⁶. Particle is a 130 nm icosahedron enclosed in a lipid envelope. Multiplication has nuclear stages.

Genus: Herpesvirus (human herpesvirus 1 (herpes simplex) and closely related viruses).
Probable genera: (No names approved.) There are many herpetoviruses of vertebrates that do not fall into the genus Herpesvirus but their classification has proved difficult. (Viruses morphologically resembling herpetoviruses have also been found in molluscs and algae.)

Family: Adenoviridae (Class I).

Double-stranded DNA of mol. wt. 20-30 × 10⁶. Particle is a 70-90 nm icosahedron which is assembled in the nucleus.

Genera: Mastadenovirus (adenoviruses of mammals).
Aviadenovirus (adenoviruses of birds).

Family: Papovaviridae (Class I).

Double-stranded circular DNA. Particles have 72 capsomeres in a skew arrangement and are assembled in the nucleus. Haemagglutinate. Oncogenic.

Genera: Papillomavirus (producing papillomas in several species of animal) 55 nm particle; DNA 5 × 10⁶ mol. wt.
Polyomavirus (found in rodents, man and other primates) 45 nm particle; DNA 3 × 10⁶ mol. wt. Includes SV40.

Family: Coronaviridae (Class IV).

Single-stranded RNA of mol. wt. 9 × 10⁶. Enveloped particles of 70-120 nm with sparse spikes. Contains a helical nucleocapsid 9 nm diameter. Cytoplasmic multiplication.

Genus: Coronavirus (avian infectious bronchitis virus and related viruses).

Family: Arenaviridae (Class V).

5-7 segments of single-stranded RNA of mol. wt. 5.5 × 10⁶. Enveloped 50-300 nm particles with spikes. Contain ribosomes and a transcriptase. Cytoplasmic multiplication.

Genus: Arenavirus (lymphocytic choriomeningitis virus and related viruses).

Family: Paramyxoviridae (Class V).

Single-stranded RNA of mol. wt. 5-8 × 10⁶ complementary to mRNA's. Enveloped 150 nm particles have spikes and contain a nucleocapsid 14-18 nm in diameter with transcriptase activity. Cytoplasmic.

Genera: Paramyxovirus (Newcastle disease virus group). Haemagglutinate; only this genus has a neuraminidase.
Morbillivirus (measles virus group). Haemagglutinate.
Pneumovirus (respiratory syncytial virus and related agents).

Family: Orthomyxoviridae (Class V).

Segments of single-stranded RNA of mol. wt. 4×10^6. Complementary to mRNA's. Enveloped 100 nm particles have spikes and contain a helical nucleocapsid 9 nm in diameter with transcriptase activity. Particles haemagglutinate and have a neuraminidase. Multiplication requires the nucleus. RNA segments in a mixed infection assort to form genetically stable hybrids within a viral species.

Genus: Influenzavirus (influenza types A and B virus).

Probable genus: (No names approved.) Influenza type C virus. Has a receptor-destroying enzyme which is not a neuraminidase.

Family: Retroviridae (Class VI).

Single-stranded 'diploid' RNA of mol. wt. $6\text{-}10 \times 10^6$. Enveloped 100 nm particles containing an icosahedral nucleocapsid. Contain RNA-dependent DNA polymerase. Only some are oncogenic. The DNA provirus is nuclear.

Subfamily: Oncovirinae (RNA tumour virus group).
          Spumavirinae (foamy viruses).
          Lentivirinae (visna/maedi virus group).

Genera: Several genera and subgenera of Oncovirinae have been defined.

VIRUSES MULTIPLYING ONLY IN INVERTEBRATES

Viruses occur not only in insects, crustacea and molluscs but probably in all groups of invertebrates. The Poxviridae, Reoviridae, Parvoviridae, Rhabdoviridae and Togaviridae (see earlier) have representatives which multiply in invertebrates. Some plant viruses are transmitted by, but do not multiply in these vectors.

Family: Baculoviridae (Class I).

Double-stranded circular DNA of mol. wt. 80×10^6. Bacilliform particles 40-70 nm $\times$ 250-400 nm with an outer membrane. May be occluded in a protein inclusion body containing usually one particle (granulosis viruses) or in a polyhedra containing many particles (polyhedrosis viruses).

Genus: Baculovirus (Bombyx mori nuclear polyhedrosis virus group).

VIRUSES MULTIPLYING ONLY IN PLANTS

Knowledge of virus multiplication is relatively rudimentary since the cell culture systems are less manageable than the animal system. Work has concentrated on physical properties and disease characteristics. Designation into families or genera has not yet been decided. Remember that the Reoviridae and Rhabdoviridae have members which multiply in both plants and invertebrates. The plant viruses listed below are not known to multiply in their invertebrate vector.

Group

Caulimovirus (Class I)-cauliflower mosaic virus group.

 Double-stranded DNA of mol. wt. 4-5 × 10^6.50 nm particles. Aphid vectors.

Geminivirus (Class II).

 Circular single-stranded DNA of mol. wt. 0.7-1 × 10^6. 18 nm quasi-isometric particles occurring in pairs and usually found in the nucleus. Persistent in whitefly or leafhopper vectors.

Bromovirus (Class IV)-brome mosaic virus group.

 4 single-stranded RNA's of mol. wt. 1.1, 1.0, 0.8 and 0.3 × 10^6. Particles 25 nm diameter. Infectivity requires 3 largest RNA's. Some with a beetle vector.

Comovirus (Class IV)-cowpea mosaic virus group.

 Two 30 nm particles containing single-stranded RNA of mol. wt. 2.3 and 1.4 × 10^6. Both needed for infectivity. Beetle vector.

Cucumovirus (Class IV)-cucumber mosaic virus group.

 Single-stranded RNA's of mol. wt. 1.1, 1.0, 0.7 and 0.3 × 10^6. 30 nm particles. Infectivity requires 3 largest RNA's. Non-persistent in aphid vector.

Nepovirus (Class IV)-tobacco ringspot virus group.

 Two 30 nm particles containing single-stranded RNA of mol. wt. 2.4 and 1.4-2.2 × 10^6. Both RNA's needed for infectivity. Some nepoviruses have a third particle containing 2 molecules of the second RNA. Nematode vectors.

Ilarvirus (Class IV)-isometric labile ringspot virus group.

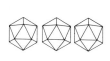 At least 3 isometric particles of different size from 26-35 nm in diameter, with 4 single-stranded RNA's of mol. wt. 1.3, 1.1, 0.8 and 0.4 × 10^6 in different particles. All RNA's, or three largest plus coat protein, required for infectivity. Seed- and pollen-borne.

Luteovirus (Class IV)-barley yellow dwarf virus group.

 Single-stranded RNA of mol. wt. 2 × 10^6. Isometric 25 nm particle. Persistent retention by aphid vectors.

Tombusvirus (Class IV)-tomato bushy stunt virus group.

 Single-stranded RNA of mol. wt. 1.5×10^6. 30 nm particle. Mode of transmission not clear.

Tymovirus (Class IV)-turnip yellow mosaic virus group.

 Single-stranded RNA of mol. wt. 2×10^6. 30 nm particle. Beetle vectors.

[?Penamovirus] (Class IV)-pea enation mosaic virus group.

 Two different sized particles (30 nm, approx.) containing single-stranded RNA of mol. wt. 1.7 and 1.4×10^6. Both needed for infectivity. Persistent in aphid vector.

[?Tobanecrovirus] (Class IV)-tobacco necrosis virus group.

 Single-stranded RNA of mol. wt. 1.5×10^6. Isometric 28 nm particle. Fungal vector.

[?Tospovirus] (Class IV)-tomato spotted wilt virus group.

 Several single-stranded RNA's with total mol. wt. around 7.4×10^6. Isometric 75 nm particles which contain lipid. Thrips vector.

Tobamovirus (Class IV)-tobacco mosaic virus group.

 Single-stranded RNA of mol. wt. 2×10^6. Straight tubular particles 300 nm long. Transmitted mechanically.

Tobravirus (Class IV)-tobacco rattle virus group.

 Two straight tubular particles. The larger is about 200 nm long with an RNA of 2.4×10^6 mol. wt. Shorter particles are of variable length and specify the coat protein. The larger RNA is infectious alone but both RNA's are needed for synthesis of new particles. Nematode vector.

[?Almovirus] (Class IV)-alfalfa mosaic virus group.

 Three bacilliform particles 18×58, 18×48, 18×36 nm and one spheroidal particle of 18 nm. Contain single-stranded RNA of mol. wt. 1.0, 0.7, 0.6 and 0.25×10^6. Infectivity needs the three largest particles, or all 4 RNA's. Non-persistent in aphid vector.

236

Potexvirus (Class IV)-potato virus X group.

Single-stranded RNA of mol. wt. 2.1×10^6. Particle is a flexuous rod with helical symmetry, 480-580 nm long and about 118S. Transmitted mechanically.

Potyvirus (Class IV)-potato virus Y group.

Single-stranded RNA. Particle is a flexuous rod with helical symmetry, 680-900 nm long and about 145S. Non-persistent in aphid vector.

Hordeivirus (Class IV)-barley stripe mosaic virus group.

Straight tubular particles with helical symmetry 110-160 nm × 20-25 nm. Three single-stranded RNA's of mol. wt. 1-1.5 $\times 10^6$ which are all probably required for infectivity. There is a variation between hordeiviruses in mol. wt. and number of RNA's.

Carlavirus (Class IV)-carnation latent virus group.

Single-stranded RNA. Particle is a flexuous rod of helical symmetry, about 690 nm long and about 165S. Non-persistent in aphid vectors.

Closterovirus (Class IV)-beet yellows virus group.

Long (600-2000 nm), very flexuous rods with helical symmetry containing single-stranded RNA of mol. wt. 2.3-4.3 × 10^6. Aphid vectors.

VIRUSES MULTIPLYING ONLY IN BACTERIA

Surprisingly little is known of the comparative biology of bacterial viruses even though molecular biology is based upon the detailed study of a few representatives. The names listed below are unofficial.

Families:
Myoviridae (Class I) (T-even phage group).

Linear double-stranded DNA of mol. wt. 120 × 10^6. Head 110 × 80 nm; complex contractile tail 110 nm long. Includes PBS1, SP8, SP50, P2.

Styloviridae (Class I) (λ-phage group).

Linear double-stranded DNA of mol. wt. 33 × 10^6. Head 60 nm diameter; long 150 nm non-contractile tail. Includes χ (chi).

Pedoviridae (Class I) (T7 and related phage groups).

Linear double-stranded DNA of mol. wt. 25 × 10^6. Head 65 nm diameter. Short tail. Includes P22.

Corticoviridae (Class I) (PM2 phage group).

Cyclic double-stranded DNA of mol. wt. 5 × 10^6, isometric 60 nm particle, lipid between protein shells, no tail. *Pseudomonas* host.

Plasmaviridae (Class I) (MV-L2 group).

ds DNA in lipid-containing envelope.

Microviridae (Class II) (φX174 group).

Cyclic single-stranded DNA of mol. wt. 1.7 × 10^6. 25-30 nm icosahedron with knobs on 12 vertices. Includes G4.

Inoviridae (Class II) (fd phage group).

Cyclic single-stranded DNA of mol. wt. 1.9 × 10^6, long flexible filamentous particle 800 × 6 nm. Host bacteria not lysed. Others infect mycoplasma. Includes M13.

Cystoviridae (Class III) (phage $\phi 6$ group).

Linear double-stranded RNA in 3 pieces of total mol. wt. 13 × 10^6. Isometric 73 nm particle with lipid envelope. Infects *Pseudomonas phaseolicola.*

Leviviridae (Class IV) (coliphage R17 group).

Linear single-stranded RNA of mol. wt. 1.2 × 10^6. 25 nm icosahedron. Includes the famous MS2 and Qβ viruses.

Suggestions for further reading

A useful source of further information is the series of volumes entitled *'Comprehensive Virology'* (Edited by H. Fraenkel-Conrat and R. R. Wagner; Plenum Press, New York). At the time of going to press there were 12 volumes in this series and more were planned. Three continuing review series, *'Advances in Virology'*, *'Progress in Medical Virology'* and *Perspectives in Virology'* are also a source of many useful articles. Some of the reviews in *'Current Topics in Microbiology and Immunology'* are devoted to aspects of virology. The major journals publishing articles on virology are:

> *Virology*
> *Journal of Virology*
> *Journal of General Virology*
> *Intervirology*
> *Archives of Virology*

Specific articles and reviews relevant to each of the chapters in this book are outlined below.

Chapter 1—Towards a definition of a virus

DANIELS C. A. (1975) Mechanisms of virus neutralization, in *Viral Immunology and Immunopathology*, ed. Notkins A. L. Ch. 5, pp. 79-97. New York, Academic Press. (A good chapter in an excellent volume.)

DELLA-PORTA A. J. & WESTAWAY E. G. (1977) A multi-hit model for the neutralization of animal viruses. *Journal of General Virology*, **38**, 1-19. (A new look at the subject with comprehensive references.)

GIBBS A. J. & HARRISON B. (1976) *Plant Virology: the Principles*. London, Arnold. (A thorough account of plant virology and by far the best book on the subject.)

HUGHES S. S. (1977) *The Virus: a History of the Concept*. London, Heinemann Educational Books. (A truly historical account of the development of our ideas about virology. Contains an interesting appendix detailing the origins of the word 'virus'.)

ROVOZZO G. C. & BURKE C. N. (1973) *A Manual of Basic Virological Techniques*. New Jersey, Prentice-Hall Inc. (Gives a detailed account of techniques in (mostly) animal virology.)

STENT G. S. (1964) *Papers on Bacterial Viruses* (2e). London, Methuen. (A collection of reprints of the original classic papers on bacteriophages. Also contains a superb introduction.)

STENT G. S. & CALENDER R. (1978) *Molecular Genetics—An Introductory Narrative* (2e). San Francisco, W. S. Freeman & Co. Ltd. (Contains much material on bacteriophages and shows how their study contributed so much to molecular biology.)

Chapter 2—The structure of viruses

CASPAR D. L. D. & KLUG A. (1962) Physical principles in the construction of regular viruses. *Cold Spring Harbor Symposium on Quantitative Biology*, **27**, 1-24. (A classic paper which discusses in great detail the principles of virus construction.)

CRICK F. H. C. & WATSON J. D. (1956) The structure of small viruses. *Nature,* **177,** 473-475. (The first exposition of the reasons why viruses may be constructed from sub-units.)

FINCH J. T. & HOLMES K. C. (1967) Structural studies of viruses. In *Methods in Virology* Vol. 3, eds. Maramorosch K. and Koprowski H. London, Academic Press. (Discusses the application of X-ray crystallography and electron microscopy to the examination of virus structure.)

KAPER J. M. (1975) *Chemical Basis of Virus Structure, Dissociation and Re-assembly.* Amsterdam, North-Holland. (Extremely detailed but an invaluable guide to this complex topic.)

Chapter 3—Viral nucleic acids

CLEMENTS J. B., CORTINA R. & WILKIE N. M. (1976) Analysis of herpesvirus DNA substructure by means of restriction endonucleases. *Journal of General Virology,* **30,** 243-256.

FIERS W., CONTRERAS R., DUERINCK F., HAEGEMAN G., ISERENTANT D., MERREGAERT J., MINJOU W., MOLEMANS F., RAEYMAEKERS A., VAN DEN BERGHE A., VOLCKAERT G. & YSEBAERT M. (1976) Complete nucleotide sequence of bacteriophage MS2 RNA: primary and secondary structure of the replicase gene. *Nature,* **260,** 500-507.

FRAENKEL-CONRAT H. (1970) *The Chemistry and Biology of Viruses.* London and New York, Academic Press. (Approximately half of this book is devoted to the chemistry of nucleic acids. Although somewhat dated it is still very readable.)

GROSS H. J., DOMDEY H., LOSSOW C., JANK P., RABA M., ALBERTY H. & SANGER H. L. (1978) Nucleotide sequence and secondary structure of potato spindle tuber viroid. *Nature,* **273,** 203-208.

SANGER F., AIR G. M., BARRELL B. G., BROWN N. L., COULSON A. R., FIDDES J. C., HUTCHISON C. A., SLOCOMBE P. M. & SMITH M. (1977) Nucleotide sequence of bacteriophage φX174. *Nature,* **265,** 687-695.

Chapter 4—Adsorption and penetration

LINDBERG A. A. (1977) Bacterial surface carbohydrate and bacteriophage adsorption. In *Surface carbohydrates of the prokaryotic cell,* ed. Sutherland I. W., Academic Press.

LONBERG-HOLM K. & PHILIPSON L. (1974) Early interactions between animal viruses and cells. In *Monographs in Virology* Vol. 9, ed. Melnick J. L. Basle, Karger. (A good review, but dating somewhat now.)

MEAGER A. & HUGHES R. C. (1977) Virus receptors. In *Receptors and Recognition* (Series A) Vol. 4, eds. Cuatracasas P. and Greaves M. F. London, Chapman and Hall.

Chapter 5—The Baltimore classification

BALTIMORE D. (1971) The expression of animal virus genomes. *Bacteriological Reviews* **35,** 235-241.

FENNER F. (1976) *Classification and Nomenclature of Viruses. Second Report of the International Committee on Taxonomy of Viruses.* Basel, S. Karger.

Chapter 6—Replication of viral DNA

GILBERT W. & DRESSLER D. (1968) DNA replication, the rolling circle model. *Cold Spring Harbor Symposium on Quantitative Biology,* **33,** 473-484. (Contains the first exposition of this controversial model for DNA replication. Many other articles in the same volume are also worth reading even though they are a bit out of date.)

KORNBERG A. (1974) *DNA Synthesis.* San Francisco, W. H. Freeman and Co. Ltd. (An excellent account of the biochemistry of DNA replication.)

LEWIN B. (1977) *Gene Expression—Vol. 3. Plasmids and Phages.* London and New York, John Wiley and Son Ltd. (Contains excellent chapters dealing with the replication of selected DNA bacteriophages.)

The reader is also recommended to read volumes 3 and 7 of *Comprehensive Virology* which deal exclusively with the replication of DNA-containing animal viruses and bacteriophages.

Chapter 7—RNA synthesis by RNA viruses

ATABEKOV J. G. (1977) Defective and satellite plant viruses. *Comprehensive Virology,* **11,** 143-200.

HUANG A. S. & BALTIMORE D. (1977) Defective interfering animal viruses. *Comprehensive Virology,* **10,** 73-116.

LEWIN B. (1977) RNA Phages. *Gene Expression,* **3,** 790-829. London and New York, John Wiley and Sons Ltd.

A review of a number of virus systems can be found in 'Control Processes in Virus Multiplication' (1975). *Society for General Microbiology Symposium* 25, eds. Burke D. C. and Russell W. C. Two chapters by Bishop D. H. L. and Flamand A.; and Burke D. C. & Russell W. C. are relevant here.

Chapter 8—The regulation of gene expression

BURKE D. C. & RUSSELL W. C. (1975) 'Control Processes in Virus Multiplication.' (This is good reading for this chapter as well as for Chapter 7.)

Chapter 9—The assembly of viruses

CASJENS S. & KING J. (1975) Virus assembly. *Annual Review of Biochemistry,* **44,** 555-611.

HERSHKO A. & FRY M. (1975) Post-translational cleavage of polypeptide chains: role in assembly. *Annual Review of Biochemistry,* **44,** 775-797.

KAPER J. M. (1975) *Chemical Basis of Virus Structure, Dissociation and Re-assembly.* Amsterdam, North-Holland. (A detailed account which contains more than you would ever wish to know.)

LEWIN B. (1977) *Gene Expression—Vol. 3. Plasmids and Phages.* London and New York, John Wiley and Son Ltd. (Contains several lucidly written chapters with a wealth of information on the assembly of bacteriophages.)

Philosophical Transactions of the Royal Society of London, Series B, **276,** 1-204 (1976) (This particular issue, number 943, is devoted to a discussion on the assembly of regular viruses. Contains excellent articles by the top men in the field!)

Chapter 10—Lysogeny

HERSHEY A. D. (ed.) (1971) *The Bacteriophage Lambda.* Cold Spring Harbor (New York), Cold Spring Harbor Press. (Seven hundred and ninety two pages devoted solely to this bacteriophage. The first 312 pages contain a series of excellent, although somewhat dated, reviews of a number of different aspects of the biology of λ. Original papers make up the remainder of the book.)

HOWE M. M. & BADE E. G. (1975) Molecular biology of bacteriophage Mμ. *Science,* **190,** 624-632 (An excellent review of this unusual phage.)

LANDY A. & ROSS W. (1977) Viral integration and excision: structure of the Lambda *att* sites. *Science,* **197,** 1147-1160.

LEWIN B. (1977) *Gene Expression—Vol. 3. Plasmids and Phages.* London and New York, John Wiley and Son. (Contains an excellent and up-to-date account of all aspects of the biology of Lambda.)

PTASHNE M., BACKMANN K., HUMAYUN M. Z., JEFFREY A., MAURER R., MEYER B. & SAUER R. T. (1976) Auto-regulation and function of a repressor in bacteriophage Lambda. *Science,* **194,** 156-161. (A lucid account of the mode of action of the λ repressor.)

Chapter 11—Interactions between viruses and eukaryotic cells

FENNER F., McAUSLAN B. R., MIMS C. A., SAMBROOK J. & WHITE D. O. (1974) *The Biology of Animal Viruses.* London, Academic Press Inc. (Still a good treatment of the whole area.)

Chapter 12—The immune system and interferon

Many of the standard texts will give sufficient background to immunology for virology.

FRIEDMAN R. M. (1977) Antiviral Activity of Interferons. *Bacteriological Reviews,* **41,** 543-567. (A recent review, but written before recent work on the mode of action.)

HOOD L. E., WEISSMAN I. L. & WOOD W. B. (1978) *Immunology.* California, The Benjamin/Cummings Publishing Co. Inc. (Detailed and up-to-date.)

HUNT T. (1978) Interferon, ds RNA and the pleiotropic Effector. *Nature,* **273**, 97-98. (A welcome review bringing order to the chaos!)

NOTKINS A. L. (1975) *Viral Immunology and Immunopathology.* New York, Academic Press. (A thoroughly recommended and excellent introduction into the complexities of this area.)

ROITT I. M. (1977) *Essential Immunology* (3e). Oxford, Blackwell Scientific Publications. (An excellent, small volume but concentrated reading.)

Chapter 13—Virus-host Interactions

An enormous area to cover, but the following provide a way in and references to further reading.

FENNER F., McAUSLAN B. R., MIMS C. A., SAMBROOK J. & WHITE D. O. (1974) *The Biology of Animal Viruses* (2e). London, Academic Press Inc. (Is still the best treatment and provides many useful references.)

GIBBS C. J. & GADJUSEK D. C. (1978) Atypical viruses as the cause of the sporadic, epidemic and familiar chronic diseases in man: slow viruses and human diseases. *Perspectives in Virology,* **10**, 161-194, ed. Pollard M. New York, Raven Press. (A recent review on a Nobel Prize winning subject.)

JAWETZ E., MELNICK J. L. & ADELBERG E. A. (1976) *Review of Medical Microbiology.* Los Altos, Lange Medical Publications. (Pp. 300-492 deal with viruses and provide a quick run-down on medical virology.)

MIMS C. A. (1976) *Pathogenesis of infectious disease.* Academic Press. (A volume from a man who has spent a lifetime working in this area.)

MORGAN E. M. & RAPP F. (1977) Measles virus and associated diseases. *Bacteriological Reviews,* **41**, 636-666. (An object lesson in the multiplicity of virus-host interactions.)

STEVENS J. G. (1975) Latent herpes simplex virus and the nervous system. *Current Topics in Microbiology and Immunology,* **70**, 31-50. (An entrée into an intriguing virus-host system.)

SMITH H. (1972) Mechanisms of virus pathogenicity. *Bacteriological Reviews,* **36**, 291-310. (Useful discussion and background.)

TER MEULEN V. & HALL W. W. (1978) Slow virus infections of the nervous system: virological, immunological and pathogenetic considerations. *Journal of General Virology,* **41**, 1-25. (Another source of recent information and references from a widely available journal.)

Chapter 14—Vaccines and chemotherapy: the prevention and treatment of viral diseases

COLLINS F. M. (1974) Vaccines and cell-mediated immunity. *Bacteriological Reviews* **38**, 371-402. (A general account of the importance of CMI.)

FINTER N. B. & BRIGDEN D. (1978) The large-scale production of human interferons and their possible uses in medicine. *Topics in Biochemical Sciences,* **3**, N76-N78. (A brief account with useful references.)

FOEGE W. H. & EDDINS D. L. (1969) Mass vaccination programmes in developing countries. *Progress in Medical Virology,* **15**, 205-243. (A discussion of the practical problems involved in vaccination in the third-world countries.)

MELNICK J. L. (1977) Viral vaccines. *Progress in Medical Virology,* **23**, 158-195. (An account of the current situation.)

Chapter 15—Tumour viruses

CAIRNS J. (1978) *Cancer, Science and Society.* San Francisco, W. H. Freeman. (Eminently readable and recommended before tackling the next references.)

HANAFUSA H. (1977) Cell transformation by RNA tumour viruses. *Comprehensive Virology,* **10**, 401-483.

TOOZE J. (1973) *The Molecular Biology of Tumour Viruses.* Cold Spring Harbor (New York), Cold Spring Habor Laboratory. (Contains numerous excellent reviews of different aspects of tumour virology, all written by experts in the field.)

Tumour viruses. *Cold Spring Harbor Symposium on Quantitative Biology,* **39**, (2 vols). (1975) Cold Spring Harbor (New York), Cold Spring Harbor Laboratory. (Detailed papers from a world-wide conference.)

Chapter 16—The evolution of viruses

BLACK F. L. (1966) Measles endemicity in insular populations: critical community size and its evolutionary implication. *Journal of Theoretical Biology*, **11**, 207-211.

FENNER F. & RATCLIFFE F. N. (1965) *Myxomatosis.* London and New York, Cambridge University Press.

JOKLIK W. K. (1974) Evolution in Viruses. *Society for General Microbiology Symposium*, **24**, 293-320. Ed. Carlile M. J. and Skehel J. J. (A general introduction dealing also with ideas on the origin of viruses.)

KAPLAN M. M. & WEBSTER R. G. (1977) The epidemiology of influenza. *Scientific American*, **237**, 88-106.

WEBSTER R. G. & LAVER W. G. (1975) Antigenic variation of influenza viruses. In *The Influenza Viruses and Influenza*, 269-314. Ed. Kilbourne E. D. New York, Academic Press Inc. (The rapidly moving area is well covered in the popular scientific press (see above) as well as the mainstream virology journals.)

Chapter 17—Trends in virology

CASALS J. (1978) New virus diseases of man. *Perspectives in Virology*, **10**, 211-233. ed. Pollard M. New York, Raven Press.

FENNER F. (1977) The Eradication of Smallpox. *Progress in Medical Virology*, **23**, 1-21.

OLD R. W. & PRIMROSE S. B. (1980) *Principles of gene manipulation: an introduction to genetic engineering.* Oxford, Blackwell Scientific Publications.

Chapter 18—The classification and nomenclature of viruses

FENNER F. (1976) *Classification and nomenclature of viruses. Second Report of the International Committee on Taxonomy of Viruses.* Basle, S. Karger.

Index